REMEMBERING CRESCENT

Remembering Crescent
Logging and Life on North Manitou Island
1907–1915

Readers are encouraged to go to www.MissionPointPress.com to contact the author, or to contact the publisher about how to buy this book in bulk at a discounted rate.

About the cover photo: First train load of logs delivered to the mill, July 12, 1909. Standing on the far left is A.J. White.

Published by Mission Point Press
2554 Chandler Rd.
Traverse City, MI 49696
(231) 421-9513
www.MissionPointPress.com

This book was funded in part by the Leelanau Historical Society's Leelanau Press Book Production Grant.

Design by Sarah Meiers

ISBN: 978-1-961302-53-2
Library of Congress Control Number: 2024907469

Printed in the United States of America

REMEMBERING CRESCENT

LOGGING AND LIFE ON NORTH MANITOU ISLAND

1907 – 1915

COMPILED BY BILLY H. AND KAREN J. ROSA

MISSION POINT PRESS

In Memory of

Esther Margaret (White) Morse

October 28, 1900–January 1, 1994

This book is dedicated in memory of my grandmother, Esther (White) Morse. Esther's childhood memories of Crescent began at the age of seven. Her family was on the island because her father, A.J. White, and brother, James, owned and operated the A.J. White and Son sawmill from 1907 until 1915. In later years, Esther took many trips with family to Lake Michigan and North Manitou, reminiscing of her island days, its people, and her beloved Lake Michigan.

— *Grandson, Billy H. Rosa*

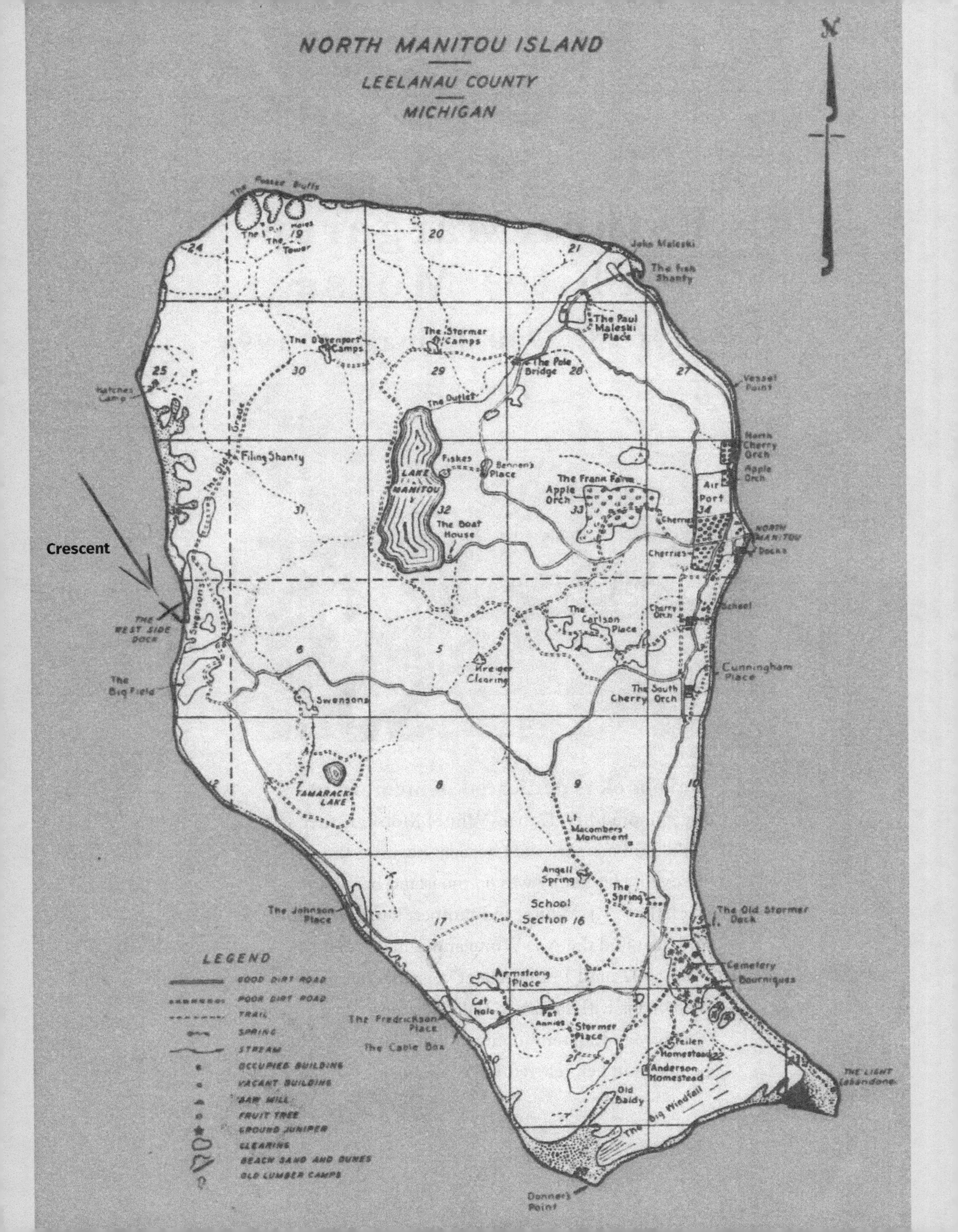

NORTH MANITOU ISLAND
LEELANAU COUNTY
MICHIGAN
N
Crescent
THE WEST SIDE DOCK
Filing Shanty
LAKE MANITOU
Fishkes
The Boat House
The Davenport Camps
The Stormer Camps
The Pole Bridge
The Outlet
John Maleski
The Fish Shanty
The Paul Maleski Place
Vessel Point
North Cherry Orch
Apple Orch
Air Port
NORTH MANITOU Docks
Cherries
The Frank Farm
Apple Orch
The Carlson Place
School
Cherry Orch
Cunningham Place
The South Cherry Orch
Kreiger Clearing
Swensons
The Big Field
Hatches Camp
The Tower
TAMARACK LAKE
Lt Macombers' Monument
Angell Spring
The Spring
School Section 16
The Old Stormer Dock
The Johnson Place
Armstrong Place
Cemetery
Bourniques
Cat hole
Stormer Place
The Fredrickson Place
The Cable Box
Feilen Homestead
Anderson Homestead
Old Baldy
The Big Windfall
THE LIGHT Abandoned
Donner's Point
LEGEND
GOOD DIRT ROAD
POOR DIRT ROAD
TRAIL
SPRING
STREAM
OCCUPIED BUILDING
VACANT BUILDING
SAW MILL
FRUIT TREE
GROUND JUNIPER
CLEARING
BEACH SAND AND DUNES
OLD LUMBER CAMPS

Important Dates

1906 Smith and Hull purchased 4,000 acres on the northwest side of North Manitou Island. They also leased additional acreage from Peter Swanson.

1907 In the fall, Monroe Dock and Dredge of Charlevoix began construction of the 600-foot dock.

November 27, 1907 The schooner *Josephine Dresden*, owned by Captain Charles Anderson, was torn from the dock and blown ashore.

1908 Captain Charles Anderson purchased the steam barge *J. S. Crouse*, and on one of the many trips, moved the White and Miser families with their belongings to Crescent.

May, 1909 The first Shay Locomotive, the No. 3, was delivered to Crescent by the steam barge *N. J. Nessen*.

July 12, 1909 The first trainload of logs was delivered to the mill. The mill was capable of producing 40,000 board feet per day.

June, 1910 The second Shay Locomotive, the No. 1, called *The Manitou Limited*, was delivered to Crescent.

January 16, 1912 J. White mill accident.

June 10, 1915 The last log was sawed at 8:30 p.m.

July 5, 1915 The White family moved off the island on the *J. S. Crouse* with the mill equipment and their belongings.

Table of Contents

Introduction

Forty years after the land was acquired by the National Park Service, the site of the village of Crescent is largely hidden in the 15,000-acre wilderness of North Manitou Island.

But the memories have been preserved. In this book, the family of Andrew James White who owned and ran the Crescent Sawmill, share their trove of historical photographs from the years 1907 to 1915.

Crescent was briefly a vibrant community, conceived and then abandoned in less than 10 years. It was a company town, built to cut, process, and ship lumber from thousands of acres on the western side of the Lake Michigan island. There were homes, barns and streets; a store, school, saloon, hotel and railroad; even sports teams. When the trees were gone, the owners shipped the machinery to other locations. The remainder was left to the forces of nature.

Within another decade, most of North Manitou Island — including the Crescent site — would come under the ownership of the Manitou Island Association, a group made up largely of Chicago-area individuals. The association continued some agriculture and timbering operations and, in 1926, brought in a small herd of white-tailed deer. From then until the 1970s, the island was mostly off-limits to residents of nearby towns on the Michigan mainland, and was known as a deer-hunting camp.

In 1970, Congress approved creation of the Sleeping Bear Dunes National Lakeshore, to include North Manitou and its smaller neighbor South Manitou Island. The National Park Service acquired North Manitou in 1984. Most of the island is now managed as wilderness, meaning no vehicles are permitted. The Crescent site can be reached on foot by trails from the Park Service dock on the east side of the island.

In the 1800s, little attention was paid to "wilderness," and Michigan's forests were valued mainly as sources of energy, building materials and chemicals. By the early 1900s, most of the prime lands had been logged off, and a remaining stand of forest on North Manitou Island had significant economic attraction.

In 1906, William C. Hull, and Franklin H. Smith purchased 4,000 acres of timberland on the northwest side of the island and made plans to harvest the timber. William was the son of Henry Hull, owner of the Oval Dish Company in Traverse City. Smith was a timber cruiser who worked for Henry Hull.

The endeavor would require a dock, sawmill, railroad, and many people. Thus, the town of CRESCENT came into being. The shoreline was in the shape of a crescent, which gave the town its name.

A.J. WHITE AND SON; OWNERS AND OPERATORS

Smith and Hull needed someone with knowledge and experience in the logging and sawmill business. A.J. White and Son had a sawmill operation in Cedar Run, Mich., at the time. A.J. White agreed to furnish sawmill equipment, oversee construction, and be in charge of the complete sawmill operation at Crescent.

A.J. White's family came to the Leelanau County community of Solon, Mich., in 1870 from Strongsville, Ohio. Andrew James was only four years old, but grew up involved in land clearing, farming, logging, and sawmill work.

A.J. owned and operated a sawmill in Cedar Run, Leelanau County, Mich., as early as 1887. According to the Manistee & North Eastern Railroad ledger, A.J. was still shipping lumber from the Cedar Run operation in 1915. Also, somewhere between 1902 and 1907, A.J. White's family, along with some friends, ventured out to Klamath Falls, Oregon, where they purchased timber acreage and built small log cabins. Within this period of time, they returned to the Cedar Run mill operation. In 1908, A.J., along with wife Clara Nell and their family, moved to Crescent. They arrived on the island on the *J.S. Crouse*, owned by Captain Anderson, and would remain on the island until July 5, 1915.

The years included some trial and tragedy. A.J. suffered a broken ankle during construction of the sawmill's millpond, and several years later had an arm amputated after an accident in the sawmill. Then, in 1914, his wife Clara Nell died from a car-train accident on the mainland. After operation ceased at Crescent in 1915, the Whites moved to Carter Siding, located north of Thompsonville in Benzie County, Mich. They owned acreage and a sawmill operation for a brief time and sold lumber and wood products to the nearby Desmond Charcoal and Chemical Plant. They also shipped lumber, lath, and shingles, on the Pere Marquette Railroad.

Prior to 1916, A.J. and son James entered into a partnership with lumbermen from New York state. Blight had affected American Chestnut trees in Virginia and other eastern states, and logging companies were salvaging the dead and dying trees. In Woodson, Va., along the Piney River, they built a new sawmill and seven miles of standard-gauge railroad into the Priest Mountains. A new hotel, a home for A.J. and Jim, and homes for loggers' and mill workers' families were also built. Along with the Whites, some families that worked at Crescent and Carter Siding also moved to Woodson to help with construction and to work in the logging and sawmill operation. In the latter part of 1919, the Whites sold all their holdings and the families moved back to Lake Ann, Michigan.

In 1920, after returning from service in World War I, James C. White joined the Tennessee Eastman Corp. in Kingsport, Tenn., as an expert in wood and timber operations. In 1926, James called his father, A.J., in Lake Ann and asked him to oversee construction of a new band-saw mill and take charge of timberland, chemical wood procurement, logging, railroad, and lumber operations. A.J. accepted and moved his family to Tennessee. In 1940, at the age of 74, A.J. retired and moved back to Lake Ann, Mich., where he continued farming and lived the remainder of his years. A.J. White passed away, Jan. 19, 1958 at the age of 92.

Esther Margaret White, to whom this book is dedicated, was born Oct. 28, 1900, in Sault Ste. Marie, Mich., and was adopted as an infant by A.J. and Clara Nell (Ferris) White. They resided at first in Solon, Mich., in Leelanau County, and moved several times throughout the lumbering days.

Those years were remembered by Esther as "wonderful childhood memories." On Oct. 12, 1919, at the farm of A.J. White, Esther became the wife of William Monroe Morse. They lived on a Benzie County farm at Almira corners adjacent to the A.J. White farm. Four children were born to Esther and William. The first was Jean Elizabeth, born July 18, 1920, mother to Billy Harmon Rosa, author of this book. Jean was delivered by Doctor Frederick Murphy, who had been the company doctor for Smith & Hull at Crescent. Esther's other children were William Julius, born Sept. 1, 1922; Helen Geraldine, born July 6, 1927; and James Kirkwood, born Jan. 9, 1929.

Esther, being the patriot that she was, offered her own personal tribute to veterans by placing American flags on veterans' graves in three Almira Township cemeteries for over half a century. William and Esther's hearts and home were always open to every need of family and community. (We could write an entire book about Esther.) William Monroe Morse passed away, March 11, 1965 at the age of 73. Esther Margaret (White) Morse passed away, Jan.1, 1994, at the age 93.

RECOLLECTIONS FROM ESTHER (WHITE) MORSE

In May of 1908, seven-year-old Esther White came on the steamship *J.S. Crouse* to her new home at Crescent, on North Manitou Island. She recalled that Dennis Robinson lifted her from the *Crouse* to a scow that took them to shore. This took place because the dock was not completed.

On one occasion in the winter, she was in need of shoes and the only shoes available at the island store were boy's shoes with hooks. She had no choice but to wear the boy's shoes or go without.

Billy Good, a cousin of Esther White's, skated from Crescent to Pyramid Point, on the Lower Peninsula mainland, to be with his wife Clara for the birth of their first child in February 1912.

At one time Esther had a toothache; Dr. Frederick Murphy came with forceps and told her he wanted to try them on to see how they fit. They fit all right and out came the tooth. Esther was mad! To make things better, Dr. Murphy had a bag of crème candy which was very special back in the day.

Esther skated for the first time on borrowed two-runner skates.

Madam Storr, a fortune teller, visited the island on May 30, 1912. She predicted that in seven years the island would sink. Esther said that "worried her." Obviously, Madam Storr was wrong.

Esther's dog "Nig" was struck by one of the logging sleighs which resulted in a broken leg. There was no veterinarian on the island so Dr. Murphy, the company doctor, set the dog's leg.

On the day the White family moved off the island, which was July 5, 1915, Lake Michigan had fairly large swells. The piano was in the hull of the boat and she remembers it rocking back and forth. The piano made the trip safely and remains in the family today.

There were times when the family would ride the *J.S. Crouse* to Traverse City. Captain Anderson would let Esther sleep on the floor between the captain's desk and the wall. She loved the rocking back and forth of the boat.

Section 1

Dock & Sawmill Construction

With thousands of acres of prime timber and ready access to major markets on Lake Michigan, the west shore of North Manitou Island seemed an ideal spot for a sawmill in the first decade of the 20th century. But before Crescent could become a functioning enterprise, a dock would be needed to bring equipment to the island and to ship finished lumber to market.

Smith & Hull Co., the project's entrepreneurs, hired the Monroe Dock and Dredge Co. of Charlevoix, Mich., to build the Crescent dock. Work commenced in 1907, and the 600-foot dock was completed the following year.

The new facility was built from wood harvested on the island and was strong enough to support railroad tracks, a steam locomotive, and huge piles of wood products. But like many of the short-lived elements of Michigan's timber era, the structures were already deteriorating 20 years after they were built.

01/01 // As construction of the Crescent Dock began in 1907, a steam engine, seen on the dock, was used to power the equipment that drove wooden pilings into the lake bed. Note the unknown woman with a child standing next to the team of horses, and the two wooden kegs and box on the dock that very well could have been full of nails and bolts. The logs in the foreground were to be used as piling.

01/02 // Construction workers stand with the wooden scaffolding during the pile-driving process on the 600-foot Crecent Dock, which was completed in the fall of 1908.

01/03 // Horse-drawn "Big Wheels" were used to haul logs to the beach for use as dock pilings. The Big Wheel was a Michigan invention that allowed timbermen to move heavy logs. Once the pilings were in place, the deck of the dock would be created from multiple layers of lumber. **// Leelanau Historical Society Museum**

01/04 // The *Josephine Dresden* was a two-masted wooden schooner built in 1853 by John Debty of Michigan. She was 95 feet long and 21 feet wide, with a capacity of 116 tons. The *Josephine Dresden* was the first vessel on the lakes to have an auxiliary gas engine, sometimes described as an auxiliary schooner or gas schooner. From 1902–1907, the schooner was owned by Capt. Charles Anderson of Sheboygan, Wis. Captain Anderson delivered equipment and material for the mill, and building supplies for the A.J. White home at Crescent.

01/05 // Charles Anderson, center, enjoys a brew with two mates on a calm day aboard the *Josephine Dresden.*

Anderson was owner and captain of the vessel until its destruction by high winds in 1907.

01/06 // On November 27, 1907, the *Josephine Dresden* was torn from the partially completed Crescent Dock and blown ashore. Capt. Charles Anderson replaced the *Josephine Dresden* with the *J. S. Crouse* and continued to provide regular service to Crescent and other ports.

01/07 // This image shows the *Josephine Dresden*, after the schooner was washed ashore but before she was taken apart by the turbulent waves of Lake Michigan. In later years, Esther (White) Morse, daughter of A.J. and Clara Nell (Ferris) White, often spoke about "we kids" playing on the remains of the *Dresden*.

01/08 // This roll-top desk was on the *Josephine Dresden* when it went ashore. The desk was salvaged before the ship came apart and given to A.J. White by Captain Charles Anderson. It remains in the family today.

01/09 // This secretary was also salvaged from the *Josephine Dresden* and given to A.J. White by Captain Charles Anderson. It is still owned by White family descendants.

01/10 // The Crescent Dock construction continued through 1907. The two masts of the *Josephine Dresden* are visible on shore in the background. On shore in front of the dock are two men with a white boat that very well could be from the Life Saving Station located on the east side of the Island. A team of horses with blankets on them can be seen close to the dock.

01/11 // Working toward completion of the 600-foot dock in 1908. On the lower left, the remains of the *Josephine Dresden* can be seen along with more logs for dock piling. The *J. S. Crouse* is visible on the far side of the partially completed dock.

01/12 // The completed Crescent Dock in the fall of 1908. Still visible in the foreground are a few remains of the *Dresden*.

01/13 // The completed dock with the railroad tracks laid in place. // **Jack Hobey Collection**

01/14 // The Shay Locomotive seen on the dock was built specifically for Smith & Hull by the Lima Locomotive and Machine Company, Lima, Ohio. It was delivered to Crescent by the Ann Arbor Railroad Car Ferry No. 4 on May 7, 1910. This locomotive was the Manitou Limited No. 1.

01/15 // This view looks toward shore from midway on the dock. The men are standing next to the railroad tracks, and the two smokestacks from the mill can be seen in the background on the left. **// Don Harrison, Collection**

01/16 // What a great view of the 600-foot dock! Note the large quantity of lumber on the dock and the vast amount of product stacked in the lumber yard waiting to be shipped. Wonder who is the brave soul standing on the ships mast? And who is the brave soul who must be standing on the ship's other mast to take the picture? **// Don Harrison, Collection**

01/17 // The large quantity of lumber on the end of the dock indicates that the mill and shipping are in full operation. The size of the waves pounding the dock offers a good indication of a strong west wind blowing across Lake Michigan.

01/18 // The wooden tramway in the lower right corner extended from the mill to the dock. Wood materials of all description are visible including, tanbark, slab wood, sawn lumber, and the logs that were used for the dock piling.

01/19 // The schooner, *Stafford*, appears to be anchored off the southwest side of the Crescent Dock while waiting to be loaded.

BILLY H. AND KAREN J. ROSA

01/20 // Another good view of the Crescent Dock, with South Manitou Island in the background. The note on the bottom of the photo says "all that is left of the *Dresden*." The two arrows depict the remains of the *Dresden*. Several people can also be seen on the dock.

01/21 // The deteriorating dock as seen in 1927. Note the layers of lumber used when building the dock. This photo was taken by Burt and Rose Gray who worked and lived on the Island during the Smith & Hull days. The big barn in the background was built about 1925–1926 by the Manitou Island Association.

01/22 // This image from 1931–33 shows the remaining pilings, while the dock boards are long gone.

01/23 // This 1937 image shows continuing deterioration of the dock piling.

01/24 // A satellite view from 2021 shows the North Manitou Island shore at the Crescent site, giving a good perspective of the size of the dock. It is remarkable how much of the submerged piling is still visible.

01/25 // The sawmill, shown during construction in 1907, was to be in operation by the fall of 1908. The workers seen are building the conveyor for the big continuous bull chain that would bring logs from the mill pond to the second story of the mill.

01/26 // This view of the mill under construction shows that the smokestacks from the boilers have been installed. Guy wires supporting the stacks are visible on close inspection.

01/27 // The image shows workers driving piling for the mill pond, which was also referred to as the "soup hole." The pond was necessary to clean the logs of sand and debris that could otherwise dull the two large circular head saws. A.J. White broke his ankle while driving the piling. The A.J. White home can be seen in the background.

01/28 // The completed mill, in operation.

A. J. WHITE & SON

PROPRIETORS

NORTH MANITOU SAW MILL

NORTH MANITOU ISLAND, MICH.

CRESCENT, MICH.,

SECTION 2
Sawmill & Crews

Steam power. Horse power. Human power.

It took a combination of raw muscle and the most up-to-date machinery of the era to convert thousands of acres of North Manitou Island forest into marketable lumber. The Crescent Sawmill was built from wood harvested on the island, but the beating heart of the operation had to be imported from the mainland.

When the mill was running, smoke from steel boilers belched out of tall metal stacks. Circular head saws were turned by a steam engine made in California. A steam locomotive from Ohio pulled rail cars along specially constructed rails.

And everywhere, there were men feeding the machinery, and horses and more men hauling, measuring, and stacking the mill's production. This was hard and dangerous work. Even the person in charge of the operation, A.J. White, twice suffered serious injuries in the Crescent Mill's eight years of operation.

02/01 // Looking from the south. Lumber sawed from North Manitou Island's maple, beech, hemlock and other species, is stacked in the yard waiting to be shipped. On the far left is the 600-foot dock extending into Lake Michigan, with another huge pile of lumber at the end of the dock. The A.J. White home can be seen on the north side of the mill next to the bluff.

02/02 // Mill crew, taken in 1911. Front row from left to right, Clyde Neufer, Lew Gibson, Jim White with the cant hook, Vern Bartholomew with the broom, Bill Bernard, Ed Gillness, Frank Mapes, Robert Dewar, and John Kabot. Back row from left to right, unknown, Edgar Youmans, John Tucker, A.J. Kidder, unknown, Rightsall, Pete Kelenske. Standing on the floor behind John Kabot is Monte White, A.J. White's brother. The sawmill workers were paid $1.75 per day, ten hours per day, six days a week.

02/03 // The four big boilers shown in the picture provided steam pressure for all the steam-powered saws and conveyors in the mill, as well as the generator that provided electric power for the mill and the town. A huge amount of slab wood and sawdust was used to fire the boilers. The man with the shovel feeding the boiler is Mont White, A.J. White's brother. In the foreground is A.J. and Mont White's nephew, Billy Good. The large boilers were delivered to the east side of the island and transported five miles across the island to the Crescent mill site by horse and wagon. Apparently, the dock at Crescent was still under construction.

02/04 // This large steam-powered engine had enough horse power to have run large and small saws, conveyors, and all other equipment. The belt on the left side of the photo is running the flyball governor, which is the vertical unit with the balls on top. This explains where the name flyball governor comes from. Governors were installed to keep the engine running at a certain RPM. The worker is unknown.

02/05 // This is the log carriage inside the mill. Alva Payne and Lew Gibson rode the carriage and set every log to maximize the most lumber. The man on the far right with his hand on the control very well could have been George Grosvenor, head sawyer. He operates the travel of the carriage with the log and men. Between the two men on the right, a large belt which powers the pulley, jack shaft, and overhead circular saw can be seen. This was called a top saw or a break-down saw and was used in conjunction with the main circular head saw on large-diameter logs. When a board comes off the carriage the man standing by the table with the rollers will send the board down to a man who runs a machine called an edger. An edger is a saw with multiple blades that saws one or both sides of a board to make it straight and correct rough-sawn width. This sawmill had the capabilities of sawing 40,000 board feet a day. **// Alva and Belle Payne (Valborg Ritola Collection)**

02/06 // The saw-sharpening room was located in or near the mill. The big circular saw is on the right and a smaller circular saw is on the left. Also shown are the shafts, pulleys, sheaves and belts that make the machinery run.

02/07 // Winter at the mill, looking from the northwest. The hotel is the building on the left. The big barn on Main Street is barely visible behind the left smoke stack. The mill pond, often called the soup hole, is at the lower left. The worker has a long pole which is referred to as a pike pole. He is using it to guide logs to the bull chain that takes them to the upper story of the mill. A pike pole can be 16 to 25 feet long with a steel tip and a curved hook. These poles were used for controlling logs floating in the water.

02/08 // December 3, 1911. Hauling slab wood and edgings from the mill to the slab pile. Note the quantity of slab wood on the left. Some of the slab wood was used to fire the mill's boilers, while some was shipped on the sailing schooners *Stevens* and *Stafford* to Wisconsin which was their home port.

02/09 // Looking from the south. The big pile of logs to the right of the building is close to the mill pond. This is an excellent view of the tramway system, with carts and wagons on the mill's upper floor loaded with lumber and ready to be transported to the lumber yard.

02/10 // Looking from the west, a wagon load of lumber ready to be hauled to the yard. Note the length of the studs on the horses' shoes to keep from slipping on the ice or wet surface of the dock and tramway. The hotel is visible in the background. Evidently, it must be wash day because there are clothes hanging on the line.

02/11 // The man on the lumber pile is Clyde Neufer and standing by the cart from left to right are Miles Jordan and an unidentified man. Asa Harvey, who appears to have a patch over his right eye, is on the horse-drawn wagon. This image again shows the studs on the horse's shoes.

02/12 // All the lumber that has been sawn arrives at this location via the "green chain." The "green chain" simply means that the continuous chain is used in transferring green, or new-sawn, lumber. The boards in the background at the far left have been transferred from the green chain onto carts and wagons to be moved to the lumber yard. Two men in the foreground are loading lumber on a wagon. To the right of the ramp are two men working on the building.

02/13 // A good view of the lower roadway and the tramway system that was used to transport lumber to the lumber yard and the dock. The power line, visible beyond the tramway, conveyed electricity from the mill's generator to various buildings in Crescent. Note Esther's dog, Nig, near the horse on the upper tramway.

02/14 // Imagine how labor-intensive a sawmill operation was in the day, when every board had to be moved by hand — often numerous times. From left to right: unknown, Miles Jordan, and Clyde Neufer.

02/15 // April 21, 1911. On the left are the hotel and Billy Good's house. The logs piled high near the center of the image are waiting to be put into the mill pond before being sawed.

02/16 // Cleaning the mill pond. Every so often it was necessary to drain the mill pond to remove and haul away the sand, bark, and debris that washed off the logs. Cleaning the logs helped prevent dulling of the big, circular head saws. Standing above the pond, from left to right: unknown, Henry Kuemin, John Tucker, Billy Good on the ladder, unknown, A.J. White, and sitting with pole in hand is Ralph Brooks. Standing in the mill pond are, from left to right, Bert Gray, unknown, Jim White with shovel, and Asa Harvey. The conveyor is located on the right with the bull chain in the middle. The bull chain moved the logs from the pond to the second story of the mill.

Esther White's recollection of her dad's accident at the mill on January 16, 1912, is as follows: A.J. White was oiling the bull chain in the second story of the mill when his heavy winter coat got caught in the chain, which mangled his left arm. Dr. Ed Thirlby and Dr. Fralick were called from Traverse City to assist Dr. Frederick Murphy, the company doctor, in the amputation of A.J.'s left arm just above the elbow. The procedure took place at the White home on the dining room table. A.J. White's instructions were to take the arm down to the mill and burn it. Esther would have been 12 years old with a vivid memory of the incident. Transportation for the doctors was as follows: They would have boarded the M&NE Railroad in Traverse City to Fouch, located close to the south end of Lake Leelanau. Then they traveled by boat to Leland where they would have boarded the mail boat to Crescent. Just imagine, no cell phones or helicopters!

02/17 // November 1909. Jim White, with the oil can, and his cousin Billy Good, holding the huge pipe wrenches, appear to be posing to create a humorous photo for Edward Beebe, the photographer.

02/18 // This image shows the mill, tramway, and loaded lumber carts.

02/19 // A mill crew, assembled at the south side of the mill, where all lumber was loaded for the lumber yard. A.J. White is at the center of the image, dressed in a suit and leaning on a post. Others identified in the group include John Tucker, fifth from left in the front row; Bert Gray, standing to the right of Mr. White; and Lew Gibson, the third man to White's right. In the back row at the far left are Billy Good and Alva Payne. The rest of the workers are unidentified.

02/20 // This style of lumber cart, with single wheel in the front, made it easy to pivot on the 12-foot-wide tramway. Note the height of the tramway. Workers could stack lumber from the ground to the tramway and just as high above the tramway, which saved valuable yard area. Remember most everything was done by hand.

02/21 // Another good view of the tramway. The man driving the horse and lumber cart is probably Asa Harvey. That would explain the notation “Go Aisy” on the picture.

02/22 // The man in the dark suit has a lumber scale stick and is measuring the “board feet” in each board that is loaded on the wagon. The stacks of lumber have narrow pieces of wood called “stickers” between the layers to allow for air circulation as the boards dry.

02/23 // Another view of the mill and the loaded lumber carts on the tramway. Note the power pole and lines coming from the mill.

CRESCENT, NORTH MANITOU.

The lumber camp of Smith & Hull has been snut down for two weeks and it doesn't look as if they intend to start up for two week's more.

The steamer J. S. Crouse arrived here this morning, after a load of horses for Peter Stormer, to return with them to Empire.

George Bernerd moved his family to Glen Haven today. We were all sorry to see them go as they will be missed by their many friends.

Mr. and Mrs. A. J. White left Friday morning for Ann Arbor, where Mrs. White has gone for medical treatment.

Mrs. Jermie Cody returned from Leland a few days ago where she has been with her little daughter. Bubie was operated on by Dr. F. E. Murphy. There was a tumor taken out of her neck. She is getting along very nicely now.

Born, to Mr. and Mrs. John Tucker, April 22nd, a girl.

Robert Dewer returned home last Friday, after spending a few days in Traverse City with his daughter, Mrs. Will Blackburn.

The young people of Crescent report a fine time at the dance last Thursday night at the hotel.

Miss Vida Collins returned to her sister, Mrs. LaCore.

CRESCENT, NORTH MANITOU.
The lumber camp of Smith & Hull has been snut down for two weeks and it doesn't look as if they intend to start up for two week's more.

:Part 1-April 1911

Traverse City Record Eagle
Traverse City, Michigan
April 26, 1911

02/24 // Mill crew on the south side of the mill. Ira Johnson is the man with the horse. Back row, seated from left to right: Clyde Neufer, unknown, Jim White, and Dave Smith. Standing in back row from left to right: unknown, Billy Good, Miles Jordan, unknown, unknown. Middle row: unknown, Henry Kuemin, Tracey Grosvenor, Pete Kelenske, Asa Harvey, Lew Gibson. Front row, seated from left to right: unknown, Walt Ramson, Bert Gray, unknown.

02/25 // The two workers pulling the wagon are A.J. Kidder on the left, and Clyde Neufer on the right. The man on the wagon is unidentified.

02/26 // April 26, 1915. This large steam engine would have been used within the mill where a significant amount of power was needed. This unit was patented by Scott & Eckhartt, and built by Union Iron Works of San Francisco, CA. The engine is 18 by 36 inches, runs at 80 rpm, and has a bed length of 21 feet. It could produce well over 100 HP, depending on the steam pressure. The belt on the right turns the pulley and the jack shaft for the flyball governor, which is the vertical unit with the balls on top at upper left.

02/27 // There is no date given for this image, but the water on the ground must be from spring thaw. The man with the pike pole, just left of center, is guiding logs from the mill pond to the bull chain that will take them up to the second story of the mill. An interesting side note: Esther White had the measles as a young girl and was confined to their house, where she had a view of the mill and mill pond. One day, through her window, she saw fire coming from the small shack next to the pond. A worker had sparked the blaze by emptying his pipe in a box of sawdust that was used for a spittoon.

TRAVERSE CITY RECORD-EAGLE Monday, December 12, 2011

LOOK BACK

News from ... 100 years ago

■ There was quite a little excitement in the little town of Crescent last week when A. J. White's barn, used by Mr. Misner, caught fire in some way. With the help of the men and hose at the mill, the fire was put out after the barn was burned quite badly.

02/28 // The mill crew, with hand carts. Top row, from left to right: Jim White, unknown, unknown. Middle row: five unknowns, Alva Payne, unknown, Billy Good, Esther's dog Nig, four unknowns. On the wagons at front, left to right: Pete Kelenske, Clyde Neufer, unknown, unknown. In the background on the far left, you can see Lake Michigan.

02/29 // The mill, from a different viewpoint. The hotel is at the far right in the background, with Billy Good's house to the left of the hotel.

02/30 // "THE BOSS" is written at the bottom of this photo of Jessie Wells. He may have been in charge of the horses, livestock, and possibly, some of the farm operation.

02/31 // Crew members are assembled in front of the men's shanty for this image by photographer Beebe. Some of the men have curry combs and brushes for grooming the horses, indicating this group may have been assigned to care for horses, livestock, and farm chores. Horses were an essential part of the lumbering operation and had to be well cared for. Seated in front with arms folded is Jessie Wells.

02/32 // July 11, 1912. An unidentified worker shows off a horse — perhaps his favorite? — in front of the men's shanty.

02/33 // Tom Parkee holds the halters of a pair of draft horses named Dick and Mag. The building behind the team is one of the barns.

02/34 // October 15, 1910. Dave Smith, on the left, is one of the blacksmiths working in the blacksmith shop. The man helping Dave is unknown. **// Photo credit: Alva and Belle Payne (Valborg Ritola Collection)**

02/35 // "RUINS OF MILL AT CRESCENT NORTH MANITOU" was written by Belle (Halvorson) Payne in the lower right corner of this photo. The last log at the Crescent Mill was sawn on June 10, 1915, at 8:30 p.m. It would take more than a year to ship away the existing stacks of lumber and to remove the railroad cars, locomotives, and other equipment from the island. In this image, the power lines are gone, boards cover holes in the roof where the smoke stacks once stood, and the Shay locomotive, Manitou Limited No. 1, can be seen just to the left of the mill. Lumber company foreman, Ed Hatch, remained at Crescent to arrange shipment for the remaining lumber. John Anderson, a long-time resident, would assist Hatch until the last load was shipped. **// Credit: Alva and Belle Payne (Valborg Ritola Collection)**

The removal of A. J. White & Sons' sawmill and the taking up of the Crescent & Southeastern railroad marks the passing of the village of Crescent, North Manitou island, Mich., which had been a hustling little lumbering town for several years.

Section 3
Logging & Lumber Camps

Talk about hard work!

In an era without gas-powered chainsaws, the lumberjacks on North Manitou Island used axes and crosscut saws as the tools to fell thousands of trees and provide a steady supply for the Crescent sawmill. Two separate logging camps were built to house the men needed to keep the operation going.

Working summers and winters for almost eight years, the logging crews kept at their jobs until all the marketable timber was cut from the Smith & Hull lands. While the tree-cutting was all manual labor, the crews used horse-drawn sleighs, "big wheels," railroad cars, and even a huge steam crane to move logs out of the forest and into the sawmill. In the final years of operation, some logs were actually rolled into Lake Michigan, floated over towards the skidway located close to the 600 foot dock, and then skidded by a steam crane up the bluff to the mill.

03/01 // A woods crew riding the rail to the logging site. In the summer of 1910, 40 Russians arrived from Chicago to work in the woods. Some of them may be among this group. The Russian workers were unskilled in logging and had a language barrier, which made things interesting to say the least.

03/02 // Teams and teamsters ready for the work day. The stump at lower left gives an idea of the size of the trees being harvested.

03/03 // It takes a lot of HORSE power to skid logs of this size.

03/04 // A set of "big wheels" in operation behind a team of draft horses. Silas Overpack, a wainwright, or wagon maker, in Manistee, Mich., is credited with inventing the big wheels in 1875, though the idea apparently came from a farmer who was one of Overpack's customers. The big wheels were used from 1875 until the 1920s. The worker standing between the wheels is Mike Jordan. He is holding a cant hook, which is a logging tool with a wooden handle and a moveable steel hook at one end called a dog. This tool is used for rolling logs and cants. A cant is a partially sawn log with at least one flat side.

03/05 // Smith & Hull bought a McGiffert large steam crane from the Clyde Iron Works Company, Duluth, Minn. It was designed to travel on the existing railroad tracks. Early on, two problems arose. The crane was too heavy for the rail system and the sandy soils. It was also slow and inefficient in retrieving logs from the woods to load on the rail cars. Later on, the McGiffert steam crane was used in skidding logs up the bluff from Lake Michigan to the millpond. We believe the McGiffert Steam Engine was delivered to Crescent by the N.J. Nessen on September 13, 1913. More explanation on photos 30–33.

03/06 // The tree in this photo will be cut down with the two-man crosscut saw, also called a "misery whip," shown in the foreground. Before cutting the tree, the men are using axes to chop a notch, which will control the direction of fall.

03/07 // The worker with the axe is Mike Jordan. The loggers have sawn halfway through the beech log with the crosscut saw. Note the steel wedge driven into the top of the saw cut to prevent the log from binding on the saw blade.

03/08 // A nice team of dapple-gray horses and a good view of how big wheels were used to move heavy logs. The leading edges of the logs were lifted so only the rear portion dragged on the ground. That kept most of the weight on the axle between the wheels, and made the job much easier for the horses.

03/09 // The log on the ground is hooked with a set of logging (skidding) tongs, often called grabs, as the team prepares to drag it from the spot where it was felled. Logging tongs were used most of the time rather than a chain which had to be completely wrapped around the log. The man on the left is Ralph Brooks.

03/10 // A well-deserved break. Note the variety of hat styles. // **Photo credit: Alva and Belle Payne (Valborg Ritola Collection)**

03/11 // A teamster proudly showing off his horses.

03/12 // These logs look like they are ready to be hauled with the big wheels.

03/13 // It is unclear what the two teams and wagons are hauling, other than people and dogs.

03/14 // This woods crew is taking a break for a photo. The image also gives an excellent view of how the standard-gauge railroad was constructed. The width between the rails is 4 feet, 8 1/2 inches. Note how far apart the wooden railroad ties are. Evidently, they are close enough to support the tracks for the short amount of time they would be needed.

03/15 // Another view of the steam crane. Note the railroad logging car and portable track that the crane could carry with it.

MEN WANTED—Sawyers, swampers and lumber haulers, at Crescent, North Manitou Island Inquire Oval Wood Dish office Smith & Hull. aug 5-tf

03/16 // Two loggers with axe and crosscut saw taking time for a photo. Pay for an unmarried lumberjack was $25.00 per month, plus meals and housing in the bunk house. A lumberjack who had a family was paid $30.00 a month, plus a noon meal and housing in Crescent.

03/17 // Logging in the winter time. Slippery snow and ice could ease the skidding of logs, but the conditions definitely were difficult for the men and the teams.

03/18 // For logging operations in deep snow, sleighs were used instead of the big wheels. Logs were loaded on the sleigh by rolling them up poles with the use of cables that were pulled by the horses.

03/19 // Three huge logs made a full load on this sleigh. On the right are more big logs, waiting for the next load. The team on the left is hooked to pull the sleigh. The team on the right would have been used to pull the cable which rolled the logs onto the sleigh.

03/20 // Another load of logs being delivered to the mill. When the loaded sleighs went down a grade, the only "brakes" were applied by letting chains drag under the runners. On the left, a huge stack of logs can be seen next to the mill pond.

03/21 // Down by the mill pond, another load of large logs is waiting to be unloaded. The A.J. White home is in the background.

03/22 // Smith & Hull contracted with two logging companies: E.W. Hatch Lumber Company and the Davenport Brothers. Both camps were located approximately two miles north of Crescent. The Hatch Camp was close to Lake Michigan, and the Davenport Camp, shown here, was located on either side of the railroad grade. Approximately one mile between Crescent and the lumber camps was a building called the "filing shanty," where presumably all the saws and axes would have been sharpened for the logging crews. There are no available photos of the Hatch Camp. Notice the wooden water tank next to the railroad tracks. Next to the small shack there are three pigs roaming.

03/23 // A closer view of the Davenport Brothers' Logging Camp buildings

03/24 // The Davenport Brothers' crew. // **Photo credit: Traverse Area District Library, Local History Collection**

03/25 // One would hope this crosscut saw had been sharpened at the filing shanty the night before. **// Photo credit: Traverse Area District Library, Local History Collection**

03/26 // Davenport Brothers' crewmen using big wheels to transport logs. **// Photo credit: Traverse Area District Library, Local History Collection**

03/27 // A good photo of a teamster and his team. // **Photo credit: Traverse Area District Library, Local History Collection**

03/28 // The team on the left is using the big wheels. The team on the right is skidding logs to the roadway so the big wheels can have access to them. // **Photo credit: Traverse Area District Library, Local History Collection**

03/29 // The logs in this image are being loaded on the Russell Railroad Car using the log boom jack. Loading logs with a boom jack is quite a task. A log would be rolled to a cable on the ground near where the farthest man to the left is standing. The cable runs up behind the man on top of the load, and to the tall boom located to the right of the rail car. It then goes through a sheave, or pulley, at the top of the boom, and down to another sheave fastened solidly to the ground. That end of the cable can be hooked onto the team of horses. When the team moves forward, it pulls the cable and rolls the log up the skid poles to the top of the stack. **// Photo credit: Traverse Area District Library, Local History Collection**

WANTED.

WANTED—Six sawyers, two ground loaders, one top loader and three teamsters to go to North Manitou island, Tuesday morning, March 15th. Transportation advanced. None but experienced workmen. Apply at Oval Wood Dish office, Smith & Hull company. March 10-3t

03/30 // 1914. Another method of getting logs to the mill was to roll them down a steep bank, called a log rollaway, and into the water where they could be floated toward the dock. Why this method? Access to the rollaway and Lake Michigan must have been closer and more efficient than hauling the logs to the railroad. Two men are standing at the top of this log pile and — look closely — two more are in the middle.

03/31 // May 15, 1915. From the rollaway some of the logs were fastened together with chains to make a big circle called a log boom. This confined the loose floating logs. The boomed logs were then towed along the shoreline of Lake Michigan toward the dock at Crescent. Farther out in the lake, the mailboat, *Lawrence*, can be seen.

03/32 // When the boomed logs reached the vicinity of the dock, they were guided to the lower end of the skidway with the help of pike poles that the workers used. In the background, a huge amount of lumber and possibly tanbark can be seen stacked on the dock, waiting to be shipped. The mailboat, *Manitou*, is also at the dock.

03/33 // The McGiffert steam crane's long cable was attached to the logs at the bottom of the skidway and used to drag them up to the landing near the mill pond. Note the huge quantity of logs that had recently been pulled up from Lake Michigan. This method is believed to have been used only in the last couple of years of Crescent's operation. On the left, next to the bluff, is the A.J. White home which now has an attached shed that was not in previous photos.

Section 4
Railroad

Horse-drawn sleighs and "big wheels" were fine for moving logs short distances, but in the age of steam the most efficient means of land transport was the railroad. The Smith & Hull company laid down miles of track into the woods and brought in two locomotives and a fleet of log-cars. The trains would load up in the woods, drop their loads at the mill pond, turn around at the "turn-table" and head right back out for more logs. The gear-driven Shay locomotives were considered ideal for logging operations on temporary tracks and rough terrain.

While the locomotives did the heavy hauling, it still took crews of men and horses to stack and secure the loads on the rail cars. Like other pieces of heavy equipment, the locomotives, rail cars, and tracks were shipped off the island after the logging and milling operations ended in 1915.

04/01 // The Shay Locomotive #3 was built Sept. 29, 1900, for C.A. Goodyear, Tomah, Wisc. In 1907, the Smith & Hull Co. acquired the #3 Locomotive. Thayer Lumber of Muskegon, Mich., had purchased 13,000 acres of stumpage in Stratford, Mich., and contracted with Smith & Hull to build a railroad into the woods and to cut and load the logs onto the Russell logging cars pulled by the Shay Locomotive #3. They were to deliver them to the Pere Marquette Railroad in Stratford for shipping to the Thayer Mill in Muskegon. On May 15, 1909, the Shay Locomotive #3 and Russell logging cars were loaded on the *N. J. Nessen* in Muskegon, and delivered to the dock at Crescent. The man sitting on the engine is Jeremy Cody.

- Stumpage is the price a private firm pays for the right to harvest timber from a given land base or acreage.
- Stratford, Mich., was located in Roscommon County, approximately 13 miles west of Higgins Lake. Stratford once was a thriving lumber town. It is now a ghost town.

04/02 // Labeled “The first carload of logs loaded on the Manitou Island, July 12, 1909, Smith & Hull,” this load is waiting to be delivered to the Crescent mill. The man standing at the far end of the logs is Ira Johnson.

04/03 // July 12, 1909: What a great day it was when the first train load of logs arrived at the mill. A.J. White is at the far left and Ira Johnson is third from the left. On the locomotive are fireman Charlie Edwards and engineer Dick Fletcher. The other men in the image are unidentified.

04/04 // This shows some of the six miles of railroad that was laid through steep and curved terrain. Still, the gear-driven Shay Locomotive had no difficulty hauling multiple railroad cars loaded with logs.

04/05 // The amount of smoke and steam coming from the locomotive's stack indicates the engine is hauling a heavy load on a steep grade.

04/06 // This Shay Locomotive, "The Manitou Limited #1," was built for the Smith & Hull Co. on November 27, 1909, by the Lima Locomotive and Machine Company, Lima, Ohio. The image shows the engine's gear-drive mechanism. The three men are unidentified.

04/07 // The Shay Locomotive, "The Manitou Limited #1," was loaded on the railroad car ferry, Ann Arbor #4, at Frankfort and delivered to the dock at Crescent on May 7, 1910. The locomotive, having just been unloaded, can be seen on the dock in the photo. The locomotive and railroad cars were leased to the Davenport Brothers, who were logging contractors for The Smith & Hull Co.

04/08 // At the controls of the Manitou Limited #1 are fireman Del Peoples, left, and engineer Joe McFadden. Prior to being engineer on the Manitou Limited #1, Joe was an engineer on the Manistee & North Eastern Railroad (M&NE) which ran from Manistee to Traverse City with multiple stops along the way.

04/09 // The Manitou Limited #1, at work. This image shows the gear drive mechanism for which Shay Locomotives are noted. These engines were especially suited to logging, mining, and industrial operations and could operate successfully on steep grades and less-than-ideal rails. The men in the engine may be Jack Edwards and Joe McFadden.

04/10 // Smith & Hull railroad tracks, curving through the woods. Note the big cut through the hill, which was necessary to keep the grade to a minimum.

04/11 // The locomotive is coming through the cut that was seen in the previous picture. This perspective shows the depth of the cut and the amount of dirt that had to be moved. This section of railroad was near the Davenport camp.

04/12 // Standing beside the Manitou Limited #1 are engineer Joe McFadden and fireman Louie Lockman. **// Photo credit: Leelanau Historical Society Museum**

04/13 // Unloading logs at the mill. The men are using cant hooks to roll the logs off the railroad logging car and into the millpond, which can be seen with logs floating in it behind the power pole.

04/14 // The Manitou Limited #1. Standing on the ground are fireman Del Peoples, left, and an unidentified engineer. On the engine, the child leaning on the jug is Hiram Ramson. His mother, Nina Ramson, is next to him and Rose Gray is at right, holding an oil can.

04/15 // This image shows the aftermath of an accident: as the #3 passed the mill pond with too much speed, it ran off the end of the track just short of hitting the mill. Note the conveyor, located above the engine, with two logs that have just been pulled from the mill pond. From left to right: holding a bottle is Philip Thiel who ran the store; looking out the window is Billy Good; seated is Jim White; and standing is engineer Dick Fletcher. The "3" on the cylinder identifies the engine. C&SE (Crescent and South Eastern) is written on the locomotive and on the fuel bunker someone has written "WRECK OF THE 3 SPOT."

04/16 // Another load of logs on its way to the mill.

04/17 // Wearing baseball uniforms in front of the A.J. White home, are Jim White, left, who was Esther White's brother, and Marvin Ferris, Esther's cousin from Tennessee. The square building with the smokestack in the background was called the roundhouse, which had a railroad turntable inside. When a locomotive was driven in, the table could be rotated 180 degrees, allowing the locomotive to move forward and back into the woods. Note the locomotive sitting beside the round house and the railroad track in front of the White home.

04/18 // The Manitou Limited #1 has a load of sawn lumber possibly headed to one of the lumber camps or a project along the grade. The engineer is possibly Tom Laird.

04/19 // Another portion of the railroad grade, with a locomotive chugging through the woods.

04/20 // Four men sitting on the Manitou Limited #1. At top are Joe McFadden, left, and Louie Lockman. Seated below: August Plamondon and Del Peoples.

04/21 // Fireman Charles Edwards is standing behind his father, engineer Jack Edwards. They were fireman and engineer from 1909 thru 1911. The man standing on the front of the engine is unknown. **// Photo credit: Empire Area Museum**

Shop Number 610 - Built for: C. A. Goodyear

Built: 09-29-1900	Class: B 25-2	Trucks: 2	Cylinders: [#-Diam x Stroke] 3 - 8 x 12
Gear Ratio: 2.176	Wheel Diam: 26"	Gauge: std.	Boiler: [Style - Diam.] W.T. - 36"
Fuel Type: Coal	Fuel Capacity: 1.75 Tons	Water Capacity: 1040 Gallons	Empty Weight: As built 43,000

Owners:
C. A. Goodyear #3 "C. A. GOODYEAR", Tomah, WI
(c1907) The Smith & Hull Co. #3, Stratford, MI
(5-15-1909) The Smith & Hull Co., *Manitou Limited RR* #3, Crescent (North Manitou Island), MI
(5-15-1909) Leased to: Davenport Bros. *Manitou Limited RI*
#3, Crescent (North Manitou Island), MI

04/22 // Shay Locomotive #3 Information. // **Photo credit: Dave Higley, Shay Locomotive History**

Shop Number 2243 - Built for: The Smith & Hull Co.

Built: 11-27-1909	Class: B 42-2	Trucks: 2	Cylinders: [#-Diam x Stroke] 3 - 10 x 12
Gear Ratio: 2.529	Wheel Diam: 29.5"	Gauge: std.	Boiler: [Style - Diam.] E.W.T. - 42.125"
Fuel Type: Wood	Fuel Capacity: 1.5 Cords	Water Capacity: 1560 Gallons	Empty Weight: As built 67,300

Owners:
The Smith & Hull Co., *Manitou Limited RR* #1 "MANITOU", Crescent (North Manitou Island), MI
(11-27-1909) Leased to: Davenport Bros. *Manitou Limited*

04/23 // "Manitou Limited #1" Information. // **Photo credit: Dave Higley, Shay Locomotive History**

WORK PROGRESSING

North Manitou Railroad Almost Completed—First Cargo of Rough Lumber Has Been Shipped.

The work on the logging road across the North Manitou island is progressing very fast and is nearly completed. The island now has mail service every day from Crescent to the camps, a contract having been let to Paul Maliska.

The cottages and hotel are beginning to fill very fast and the people of the island look for the biggest resort season in the history of the island.

The first cargo of rough lumber was shipped last Monday. The steamer Fletcher of the Mission line carrying about 300,000 feet to Chicago.

SECTION 5

Ships & Shipping

Crescent's short life-span, in the first two decades of the 20th century, was near the end of a long transition in the Great Lakes maritime industry. Many pure sailing vessels were still in service, but they were gradually being replaced by steamships and by vessels that took advantage of both sails and steam engines. Some smaller boats were already using gasoline or Diesel motors.

Northern Lake Michigan in those days was a busy highway with ships hauling passengers and cargo literally in all directions. As an island town, Crescent was heavily reliant on shipping, and residents would have been closely aware of arrivals and departures at the dock.

A lighthouse and a lifesaving station were in operation on North Manitou Island at the time, though both have long since been decommissioned. The lighthouse buildings are gone, while the National Park Service has offices in the former lifesaving station.

05/01 // The *N. J. Nessen* is tied to the Crescent Dock and taking on a load of lumber. Note the stacks of boards already loaded. The *N. J. Nessen* was built in Lorain, Ohio, in 1880, and was home-ported in Grand Haven. It was 148.6 feet long and 37 feet wide, with capacity of 440 gross tons. With two masts and a smoke stack, the *Nessen* is technically called a steam screw, or steam barge.

05/02 // At the Crescent Dock. Slabs and edgings are being unloaded from wagons and onto one of the schooners, most likely the *Stevens* or the *Stafford.*

05/03 // The steamer, *J. S. Crouse*, at the Crescent Dock. The *J. S. Crouse* was purchased by Captain Charles Anderson in 1908, after the wreck of his schooner, the *Josephine Dresden*. The *J. S. Crouse* was built in 1898, by James Elliot in Saugatuck, Mich. Gross tonnage was 82.26. It was called a steam screw, or steam barge. From *Grand Traverse Herald, June 2, 1910: "The steam barge, J. S. Crouse, delivered a load of hay and grain to Crescent for Smith and Hull."* On November 15, 1919, the *Crouse* caught fire and was run ashore 500 feet from the Glen Haven dock, where it was destroyed. The cargo at that time was lumber and potatoes from Milwaukee.

Str. Seneca aground N Manitou 1911

POST CARD

048
748

CORRESPONDENCE HERE

This is the "Seneca" the boat that went ashore. I took this from the top of our mill. Mother.

NAME AND ADDRESS HERE

U.S. POSTAGE ONE CENT

DEC 22 1911 MICH

Geo. G. Thirlby
204 E. 8th St
Traverse City
Mich

05/05 // The previous photo of the *Seneca* was taken from the top of the saw mill by Clara Nell White (Mrs. A.J. White). This postcard with the message was sent to George & Nellie Thirlby, the Whites' daughter and son-in-law, on December 22, 1911.

05/04 // On November 11, 1911, the *Seneca* ran aground just west of the Crescent Dock. The four-mast steamer made of steel was built in 1889, in Cleveland, Ohio, by the Globe Iron Works. The length was 290 feet; width, 40.66 feet.

SENECA IS ASHORE

LEHIGH VALLEY STEAMER IN DANGEROUS SITUATION.

Aground On West End of North Manitou Island In Badly Exposed Position.

Glen Haven, Mich., Nov. 23.—The Lehigh Valley steamer Seneca is ashore on the west end of North Manitou island. She is cut one foot forward and three foot aft. The vessel is resting easy at present, but she is in a badly exposed position. The wind is changing to northwest, making her position a critical one and very dangerous.

Traverse City Record Eagle
November 23, 1911

05/06 // Traverse City Newspaper Article, December 23, 1911.

05/07 // The *Lawrence,* operated and possibly owned by Joe Haas, was one of the mailboats that delivered mail from Leland to the east side of North Manitou Island. Judging from the photo, it very well may also have transported passengers. An article in the *Leelanau Enterprise*, on Thursday, July 11, 1912, reported a tragedy: "*Joseph Haas of South Manitou drowned yesterday while lifting his fish nets.*"

05/08 // The *Manitou* was also used as a mailboat that delivered mail to the Island and very well could have transported passengers. John Paetschow operated and possibly owned the *Manitou*.

05/09 // After arriving at the east side of North Manitou, mail was driven by horse and buggy to Crescent, on the west side. Esther White recalled that the cross-island delivery was accomplished by Paul Maleski, Sr. Paul is leaning here on a capstan, a vertical, rotating unit used on land and on ships for winding ropes and cables to secure the vessel.

05/10 // The men on the right are unloading and stacking lumber that had just been delivered from the lumber yard to the Crescent Dock. The men on the left are leaning on a vast amount of large timbers, possibly railroad ties. Also, note the schooner anchored off the west side of the Crescent Dock. **// Photo credit: Jack Hobey Collection**

05/11 // The *F. W. Fletcher* is at the end of the Crescent Dock, being loaded with lumber or possibly tanbark. The *F. W. Fletcher* was built in 1891 in Marine City, Ind., and is called a steamer, or steam screw. Length, 161 feet; width, 32 feet; depth, 11.25 feet; gross tonnage, 495.95; net tonnage, 314.13 ton. The hull is wood. The steam ship in the background is the *J. S. Crouse*.

CRESCENT.

The barge Fletcher took on a cargo of lumber here today.

A social given by the school Saturday evening was well attended, the proceeds amounting to $31 dollars, to be used for school library.

Mr. and Mrs. A. F. White and family will spend the following week with friends in Traverse City, Solon and Cedar Run.

Miss Belle Halverston is planning on a week's trip to her home in Solon.

Mrs. Ashley arrived last evening, and they will soon be settled in their home here.

Miss Bertha Thiel expects to visit her parents soon.

We have had no mail for the past week on account of the wind storm.

Mrs. Margrette Yoemens and son were calling on Mrs. Kimbaell Saturday afternoon.

Miss Ada Brown is helping Mrs. Ransom the past week.

05/12 // The sailing vessel, *Stafford*, is seen off the south west side of the Crescent Dock, in unusually calm Lake Michigan waters.

05/13 // This photo shows the *Stafford*, with a portion of her sails rigged, closer in toward the dock. The *Stafford* was built in Tonawanda, NY, in 1868. It was 112 feet long and had 190-ton capacity.

05/14 // The *Stafford*, having been loaded with tanbark, is anchored off the northeast side of the Crescent Dock trying to withstand the turbulent Lake Michigan waters. Note the large quantity of lumber stacked on the dock.

05/15 // This photo is the first in a sequence taken as the *Stafford's* crew and men from the U.S. Lifesaving Service respond to an emergency that occurred as the *Stafford* broke free from the dock and was washed toward shore.

05/16 // The second photo in the sequence shows the *Stafford*, with its load of tan bark, drifting toward shore in stormy waters.

05/17 // As part of the same sequence, crewmen in a boat from the North Manitou Island Lifesaving Station make their best efforts to deal with the *Stafford* situation.

05/18 // Here, in the final image of the sequence, the storm appears to have substantially subsided. The men have removed their rain gear and are surveying the situation. (This photo and the previous four were taken around 1910 and 1911.) The image is also a good view of the tan-bark cargo. For a long time, this was considered a worthless waste product. But it was discovered that the bark of the hemlock, otherwise known as tanbark, contained tannic acid which could be extracted and used for tanning leathers.

05/19 // The *F. W. Fletcher*, on the northwest side of the Crescent Dock, is being loaded with lumber. Again, note the huge amount of lumber stacked on the dock.

05/20 // This image shows the *F. W. Fletcher* (the same ship as in the previous photo) as viewed from shore looking at the ship's bow.

05/21 // Loading lumber from the Crescent Dock onto the steam screw, the *Helen Taylor*. The *Helen Taylor* was built in 1894 in Grand Haven, Mich. From left to right: Esther White, Rose Gray and Mrs. Dave Smith.

05/22 // At the end of the Crescent Dock, onlookers are watching whatever is happening with the two sailing vessels. The girl with the ribbons in her hair is Esther White. The *Stafford* is the ship closest to the dock, loaded with what appears to be slab wood and with sails set. The ship at far left, with the line attached and no sails set, is the *Stevens*.

05/23 // Simply a beautiful photo of a sailing ship anchored offshore.

05/24 // The motor vessel *Hurry Back*, berthed onshore near the Crescent Dock.

An excerpt from the story of the *Hurry Back* as told by Enus Swanson, the son of Peter Swanson:

In 1908, Pat Huey wanted to become partners with Peter Swanson in the fishing business. They installed a twelve horsepower Duefree motor that Pat had purchased and Pat & Peter built a cabin on the twenty-five foot long, nine foot wide boat which had previously been a sailboat. On Tuesday, October 20, 1908, Peter, Enus, Pat Huey, and his wife left Crescent in the fishing boat the *Hurry Back* because trout season had opened October 1st. The first nets were set about 1:00 p.m. and the next morning they would lift the nets and set more. Some cutover land was on fire on the eastern side of the state and a strong wind came from the east covering Lake Michigan with smoke. It was thick like fog and you could not see far. "Well they started for shore that we didn't find." "We should be back by 4:00 but was not." A storm came up and the motor quit and they went where the wind carried them. They drifted to Detroit Island located in Door County, Wisconsin. They spent Thursday and Friday on the island with very little food. On Friday night they were rescued by the crew from the Life Saving Station from Plum Island having seen the big fire Peter Swanson had built on the shore. Saturday and Sunday was spent repairing and getting the engine to run on the *Hurry Back*, and contacting the Life Saving Station on North Manitou to accompany the *Hurry Back* to Crescent. Oscar Smith and Ed Fisher left the Life Saving Station on North Manitou at 4:00 a.m. on Monday and arrived at Plum Island at 10:30 a.m. At 1:00 p.m. They were all headed for Crescent. The engine on the *Hurry Back* worked for awhile and then quit. "It was up to Oscar Smith to tow the thing to Crescent." "We landed at Crescent." "The beach was all lit up, everyone was there with a lantern in their hand, and had a fire on the beach, too."

Enus's note about the storm.

"The storm had been very bad and we had been reported gone for good." "All boats on the lake big and small were in harbor or in shelter for the storm." "Except 2 lumber barges loaded with lumber was missing, crew and all was gone." "They said that some of the lumber should hit North Manitou."

North Manitou Island Light was a Lake Michigan lighthouse and fog signal complex at Dimmick's Point on North Manitou Island in Leelanau County within the U.S. state of Michigan. In operation from 1899 until 1935, the lighthouse helped to mark the Manitou Passage. The lighthouse complex was superseded by the North Manitou Shoal Light in 1935 and was privatized in 1938. While fragments of the light station complex remained as of 2017,[1] the lighthouse tower succumbed to the effects of erosion on its exposed site in 1942. The lightkeeper's dwelling house followed it into oblivion in the early 1970s.[2]

05/25 // Information about the North Manitou Island Lighthouse.

05/26 // The North Manitou Lighthouse complex. The large brick building is the lighthouse keeper's residence. The building in the foreground could have been used for buggies, horses, and other equipment. The building to the left with the two stacks is the foghorn building. Barely visible out on the lake are a three-mast schooner to the left of the lighthouse, and a large steam barge to the right. Both ships appear to be going through the Manitou Passage.

05/27 // Another view of the lighthouse. A steamer can be seen to the left of the tower, far out on the lake. The brick building with the two stacks is the foghorn building. It contained two wood-fired boilers which produced steam power for the two foghorns. Only one foghorn was used at a time and the other was a backup. Note the huge piles of firewood used to fire the foghorn boilers and possibly lighthouse keeper's residence. The small brick building in the foreground may have been used to store the kerosene which lit the lighthouse lamp.

05/28 // The lighthouse keeper's residence, shown from the front. The American flag is displayed at the top of the flag pole, with nautical signals on either side. Lighthouses and signal flags were essential navigation aids that helped ships at sea in different ways.

05/29 // The North Manitou Island Lighthouse Keepers quarters, from a slightly different angle.

Section 6
School & Students

Lumber boomtowns, in the common image, were mostly populated by hard-working and probably hard-living men.

There was some of that in Crescent, to be sure. But there was also a softer side as seen in the community's efforts to employ teachers and provide an education for the children of families in the town. Interestingly, a former saloon was converted into the school — which also hosted church services. Esther White attended the one-room schoolhouse from age 8 to age 15, when her family moved off the island.

06/01 // Esther White recalled that the first school sessions in Crescent were held in a log building, and that the 1908 school "year" was only four months long. The first teacher was Esther's cousin, Mary Cate.

Left to right, in back row: teacher Mary Cate, Enus Swanson, Ada Brown, Viola Miser, Beulah Miser, Lillian Bernard.

Middle row: Unknown, Louie Bernard.

Front row: Esther White, Mary Podleski, Thelma Kidder, Lena Bernard, Genevieve Bernard, Lynn Bernard, Unknown.

06/02 // Esther White said the building in this photo was originally the saloon, which shortly thereafter became the schoolhouse. This building also served as the church and Sunday school. Note the power lines going to the building. The school photos are believed to be from 1908 through 1911.

06/03 // This photo was taken in front of the one-room school house.

Back row from left to right: Viola Miser, Ellen Gibson, teacher, Belle Halvorson.

Middle row from left to right: Ralph Gibson, Eddie Wyse, Layman Burkette, Louie Bernard, Tracy Grosvenor.

Front row from left to right: Earl Burkette, Esther White, Alice Gibson, Lena Benard, Mary Podleski, Thelma Kidder, Lynn Bernard, Earl Gibson, Ada Burkette, Beulah Miser.

06/04 // No information or pictures are available regarding the Crescent School, beyond the years of 1908–1911. It is known that Vera (Wynkoop) White, Jim White's wife, was also a teacher during the Crescent years. The number of young people in this group indicates that the one-room school must have been a busy place.

Back row from left to right: Ellen Gibson, Belle Halvorson, Frank Podleski, Beulah Miser, Viola Miser, Louie Bernard, Ralph Gibson.

Middle row from left to right: Unknown, Lena Bernard, Esther White, Tillie Tucker, Earl Gibson.

Front row from left to right: Uknown, Unknown, Unknown, Unknown, Edna Tucker, Alice Gibson, Unknown, Unknown, Lynn Bernard.

SECTION 7

Baseball & Basketball Teams

A sports team can help create a sense of identity in a community — even when the community itself is only a few years old. As the images in this section testify, the little town of Crescent boasted at least three baseball or basketball teams during the town's eight years of existence.

In addition to giving young men and women a break from the routine of life in an isolated lumber town, the games provided opportunities to exercise bragging rights over nearby communities on the South Island or the mainland.

07/01 // In the early 1900s, it was common for towns and villages to have ball teams. Crescent was no exception. The town had two baseball teams. One was called the Tigers. A Native American team was called the Red Wings. Crescent also had a girls' basketball team.

The 1911 team, from left to right: Jim White, Roy Beeman, Bramer, John Winters, Del Peeples, Coach Dr. Otto M. LaCore, Louie Lockman, Alva Payne, Asa Harvey, Rightsall, Ira Johnson.

07/02 // Jim White, left, and Ralph Brooks in an image from 1910.

There was a big ball game on Sunday, April 23rd, between the mill crew and camp crew. The line-up was as follows:

Smith & Hull Camp crew—Pontiac, ss.; Smith, 1b; Beolles, 3b; Winters, p; Harton, 2b; Freeman, lf; Roxbury, c; Ferney, rf; Willard, cf.

A. J. White, Mill crew—Payne, 2b; Brooks, 3b; Ferris, ss; White, c; Jordan, lf; Podliski, 1b; Gray, cf; Grovener, rf.

Score by innings:

Mill Crew ....3 1 2 0 0 0 4 1 x—12
Camp Crew .....0 0 1 0 0 0 0 0 0— 1

Runs—Payne 2, Brooks 3, Ferris 1, White 3, Jordan 2, Gray 1. Two base hits — Rosburry, Pontiac, Payne, Brooks, White. Sacrifice hits—Brooks. Left on bases—Mill crew 8. Camp crew 6. Stolen bases—White, Jordan, Payne, Brooks, Grey, Pontiac. Strick out by Winters 11, by Jordan 13. Bases on balls off Winters 4, off Jordan 1. Home run by Jordan. Hit by ball, by Winters, Jordan. Time 1 hour, 40 minutes. Umpire, O. M. LaCore.

April 25.

Crescent Part 2-April 1911

Traverse City Record Eagle
Traverse City, Michigan
April 26, 1911

07/03 // Cary Hull of Smith & Hull made a trip to North Manitou in September, 1910, to evaluate the progress of the lumbering operations at Crescent. The ball team was having practice and encouraged Mr. Hull to join in the game.

07/04 // This photo is of the Native American team called the Red Wings. Most of the players were from Peshawbestown, in Leelanau County between Suttons Bay and Omena, and likely were members of the Grand Traverse Band of Ottawa and Chippewa Indians.

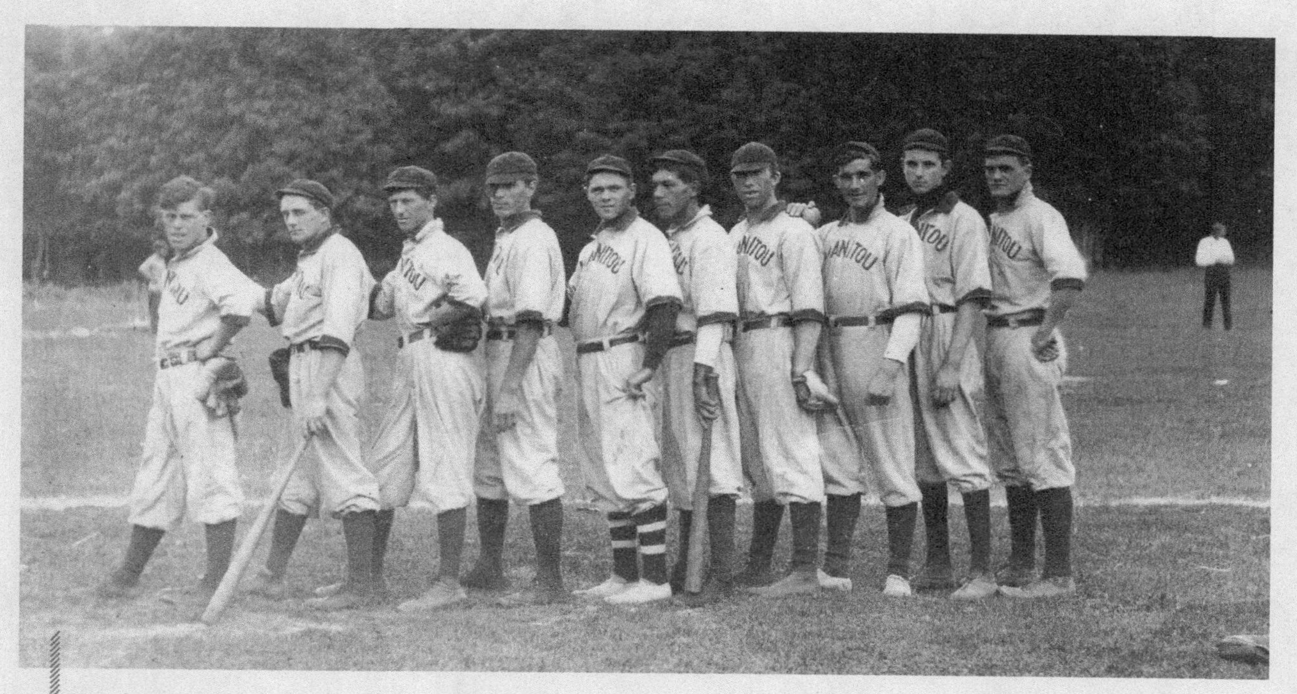

07/05 // From left to right: Ira Johnson, Ralph Brooks, Alva Payne, Asa Harvey, Roy Beeman, Johnnie Jingway, Miles Jordan, Unknown, Marvin Ferris, Jim White.
Leelanau Enterprise, July 11, 1912: "The baseball game between Cedar and North Manitou teams on the 4th resulted in a score of 11 to 4 in favor of North Manitou.

07/06 // (This photo and the previous one are from the same day)

Back Row from left to right: Roy Beeman, Unknown, Jim White, Marvin Ferris, Miles Jordan.

Seated from left to right: Asa Harvey, Ira Johnson, Johnnie Jingway, Ralph Brooks, Alva Payne.

07/10 // John Winters and Jim White. Jim is holding the catcher's mitt, and the mask is at his feet.

07/07 // This image from 1911 indicates the ladies also had a ball team. Basketball, that is.

From left to right: Eva Swanson, Ellen Gibson, Beulah Miser, Ada Brown, Viola Miser, Coach Belle Halvorson, Laura Peterson, Rose Gray, Nina Ramson, Alice Turner, and Belva Grosvenor.

07/08 // Another photo of the girls' basketball team.

Standing from left to right: Eva Swanson, Viola Miser, Beulah Miser, Coach Belle Halvorson, Belva Grosvenor, Laura Peterson, Rose Gray. Seated from left to right: Ellen Gibson, Ada Brown, Nina Ramson, Alice Turner.

1909 Escanaba Girls Basketball team

Special to the Record-Eagle/Benzie Area Historical Museum

The 1911 Benzonia Academy girls basketball team. The academy, and the colleges that proceeded it, was built on the idea of an academic education for all, regardless of sex or race.

07/09 // Rose Gray, with basketball marked "1914."

Though no information could be found regarding the girls basketball games, other sources confirm that several Northern Michigan communities also sponsored girls' teams during this period.

07/11 // June 12, 1913. Back row from left to right: Roy Beeman, Miles Jordan, Jim White, Earl Daily, Clyde Neufer. Seated from left to right: Ed Cate (cousin to Jim White), Ralph Brooks, Johnnie Jingway, Ira Johnson.

Leelanau Enterprise, June 19, 1913: "Captain John Paetschaw brought the Leland Baseball Team to Crescent Sunday last (June 12, 1913) that resulted in a score of 6 to 10 in favor of Leland."

07/12 // June 12, 1913 (same day as previous image):

From left: Ed Cate, Ralph Brooks, Jim White, Ira Johnson, Miles Jordan, Roy Beeman, Lockman, Clyde Neufer, Johnnie Jingway. **Perhaps the cattle in the background are keeping the outfield mowed and fertilized!**

07/13 // Miles Jordan sporting his "Crescent Tigers" baseball uniform.

07/14 // Another good photo of the Native American team, the Red Wings. Unfortunately, we do not have information on the players' names.

07/15 // In a sequence from what appears to be a practice session on a less-than-groomed field, the top image shows Miles Jordan pitching. The lower photos show, from left, Alva Payne, Ralph Brooks, and Miles Jordan.

SECTION 8
Town & Families

Crescent was built as a business enterprise. It existed for the sawmill and lumber camps that were created to extract value from the North Manitou Island forest. As in other lumber towns, buildings were thrown up quickly, and fell apart almost as soon as the timber was gone.

And yet, in its eight years of existence, Crescent had many of the same characteristics as any small town at the end of the horse-and-buggy era. The main street had a store, post office, school, homes, and, of course, horse barns. Overhead wires supplied electricity. Men found time for sports teams, billiards, and other distractions when they weren't at work. Young people fell in love, married, and had babies. When the island adventure ended, the families moved away to new homes. And they carried memories of Crescent for the rest of their lives.

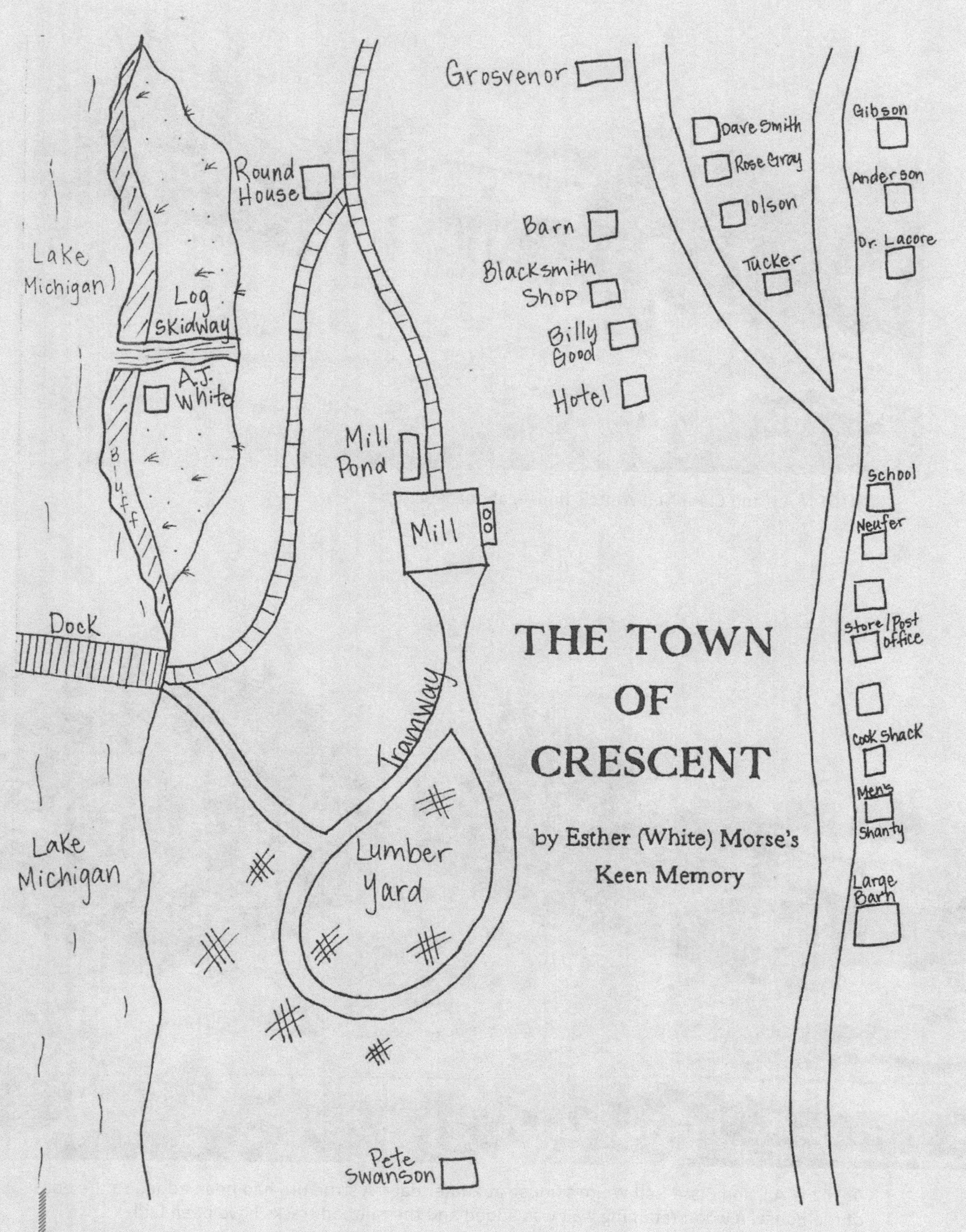

08/00 // Map of Crescent, sketched by Emma Rosa from Esther (White) Morse's keen memory.

08/01 // A.J. and Clara Nell White's house, about 1908.

08/02 // A.J. and Clara Nell White's house at a later date. A structure had been added to the end of the house. A wood retaining wall was added and the railroad tracks have been laid.

08/03 // A.J. White and Clara Nell (Ferris) White, March 18, 1887.

08/03.1 // A.J. White, grandson James Thirlby, and son, James White.

08/04 // From left: Clara Nell White, Esther White, Alice Turner, and Laura Peterson.

08/06 // Esther White, 10 years old, standing, and Pearl Miser 1½ years old.

POST CARD

CORRESPONDENCE HERE

NAME AND ADDRESS HERE

CRESCENT MAY 17 1912

I haven't time to write a letter. I am so glad Geo. is better. I am feeling fine. The Stafford is loading slabs today 176 cords, the Crouse and Stevens both loaded this week. Pretty good. Mama

Mrs. Geo. Thirlby
204 E. 8th St
Traverse City
Mich.

08/06 // A postcard image of Nellie and George Thirlby. Nellie White and George Thirlby were married at the home of the bride's parents, Mr. and Mrs. A.J. White, at Crescent in 1908. The minister, A.A. Allington, from the Oak Park Congregational Church in Traverse City, came to the island to perform the ceremony. George Thirlby had had health issues since his return from the Spanish American War in 1898; he ultimately died of malaria on January 22, 1913. Many said that Nellie Adelia White, who died three months after her husband, died of a broken heart. She knew her mother would care for her young son, James.

08/07 // This message was on the back of the previous postcard photo, and was sent from Mrs. A.J. White to daughter, Nellie (White) Thirlby. Note the information about the ships that had been loaded that week.

08/08 // Nellie (White) Thirlby, George Thirlby, and son James, born August 3, 1910.

08/09 // Nellie and James Thirlby taken on the porch of Nellie's Aunt and Uncle's home, Bertha and Mont White.

08/10 // James Thirlby, 3 years old, and Esther White, 13 years old, on the steps of A.J. White's home.

08/11 // Viola, daughter of Eli & Carrie Miser. On August 24, 1919, in Northport, Viola married Robert (Bob) White of Solon, son of Thomas R. and Minnie (Baine) White. Bob was the nephew of A.J. White. Bob White served as sheriff of Leelanau County from 1938 to 1964.

08/12 // Beulah Miser holding younger sister Pearl. They were daughters of Eli and Carrie Miser.

08/13 // The Hatches west side hotel at Crescent. The hotel was run by the Eli and Carrie Miser Family.

08/14 // 1912, on the hotel porch.

08/15 // Playing cards: One of the great pastimes that took place in the hotel.

08/16 // Not only are cards played inside the hotel, but outside as well.

08/18 // Dave Smith, his wife, and their three girls on the floor of their new house, which is under construction.

08/19 // The Smith girls, all dressed up.

08/20 // Lou Gibson's wife and children. In front are sons Ralph and Earl, and daughter, Alice. In back are Lou's daughter, Ellen, and wife, Margaret. Lou is not in the photo. He works in the mill, riding the log carriage. // **Photo credit: Traverse Area District Library, Local History Collection**

08/21 // The barber shop at Crescent. Through the doorway, a patron can be seen in the barber chair.

08/22 // The A.J. Kidder family. Standing, from left, are: A.J. Kidder, holding George; his wife, Ida; and daughter Thelma. Seated left to right are Bernice and Frances. Note the row of workers' houses in the background.

08/23 // This photo of thirsty lumberman would indicate the building they are standing in front of is the saloon, prior to becoming the school house. **// Photo credit: Leelanau Historical Society Museum**

08/25 // 1907–1908. This photo shows the first buildings constructed on the main street. The barn is in the foreground, the building with the cupola is the men's shanty, and the third building is the cook's shack.

08/26 // A closer view of the men's shanty and the cook's shack. At this early stage of development, no other buildings have been built to the left of the cook's shack.

08/27 // As the town develops, on the right a new gambrel-roofed barn has been built. Seated in the buggy are Mary Swanson and daughter, Eva. Also, note more power poles, power lines, and buildings on the main street.

08/28 // Another photo of the main street. Note the dinner bell on the pole at the cook's shack. The building with the covered porch is the store and the post office.

08/29 // The view of the main street from the opposite end, with the school house in the foreground.

08/30 // Lumberjacks enjoying some free time playing pool. Note the one electric light above the pool table and the style of support used where the two beams are joined.

08/31 // Another game of pool.

08/32 // This duplex was the home of Mr. & Mrs. Robert DeWar, on the left, and Mont & Bertha (Thiel) White and daughter Mary, on the right.

08/33 // Bertha White and baby daughter Mary, at home.

08/34 // Mont White, left, and Robert DeWar, in 1908. Bet that could have been interesting!

08/35 // 1910. Walter, Nina and Hiram Ramson. Note that Esther's dog, "Nig," also got into the picture.

08/36 // The Ramson family, about 1914. Walter worked a few months for Capt. Charles Anderson on the steamer, *J. S. Crouse*, and then went back to work at the mill.

08/37 // The Swanson family: Mary, Peter, son Enus, and daughter, Eva.

08/38 // Peter and Mary Swanson's home, with Mary overlooking a variety of knick-knacks that she may have been collecting and/or selling. Printing on the upside-down box just left of center reads: MAPLE CITY CREAMERY, BUTTER MADE FROM PURE PASTURIZED CREAM. MAPLE CITY, MI.

08/39 // Peter, Mary and Eva Swanson, riding in the buggy. Note the Big Wheels for logging in the background and, to the left, a large pile of slab wood. **// Photo credit: Jack Hobey Collection**

08/40 // Peter Swanson, stopping a minute for a photo.

08/41 // The return from a successful fishing trip. At left is Peter Swanson, holding a nice fish. John Holt is in the middle. Peter Swanson's son, Enus, is in the bow of the boat, holding another fish. More of the catch is in the boat. Note all the slab wood in the background.

08/42 // The Smith & Hull General Store and Post Office. The store was run by Phillip and Jennie Thiel who are in the photo. The first postmaster was Frank Dean, September 21, 1908.

08/44 // Donald, son of Philip and Jennie Thiel, with his loyal companion.

08/43 // Inside the Smith & Hull store. Note the rack standing on the counter of the store holding many Beebe postcards for sale.

08/45 // In the buckboard at the front of the store and post office are Mary Swanson and daughter Eva. Standing on the left near the horse is Dr. Otto M. LaCore who replaced Dr. Frederick Murphy M.D. as the company doctor. On the right is Earl Dailey. On July 14, 1910, Phillip Thiel became the postmaster. The post office and general store closed August 31, 1915.

08/46 // Ada (Brown) Jordan and Miles Jordan leaning on the fence in front of their home. They were married in Crescent on June 19, 1911.

08/47 // Ada (Brown) Jordan and her mother, Hattie, holding Ada's first baby. The calendar on the wall is dated 1912.

08/75 // Miles Jordan is holding the chicken. On the right is his wife, Ada (Brown) Jordan. In the middle is Ada's mother, Hattie Brown. // **Photo credit: Jack Hobey Collection**

08/48 // The two girls in the photo are Edna Tucker, left, and Tillie Tucker, daughters of John and Cora Tucker of Lake Ann. The Crescent news column in the *Traverse City Record-Eagle* on April 26, 1911, stated that on April 22, 1911, a daughter, Beatrice, was born to John and Cora Tucker.

08/49 // Alva Payne of Cedar Run.

08/50 // Belle (Halverson) Payne.

08/51 // These are Belle & Alva Payne's children, Valborg and Carol. The rest of the story is written on the photo by Bella Payne. // **Photo credit: Alva and Belle Payne (Valborg Ritola Collection)**

08/52 // One could assume that this cat and kittens are Valborg and Carol's pets. // **Photo credit: Alva and Belle Payne (Valborg Ritola Collection)**

08/53 // The cooks and kitchen help, taking time out for some fresh air and a photo. Seated on the box is Tracy Grosvenor. The others are unknown.

08/54 // Inside the cook shack. From left to right: Tracy Grosvenor, Bramer, and unknown. Note the size of the kettles. On the far right, on the floor, is a large container that says "Armour Pure Lard."

08/55 // Must be the day to butcher hogs!! Obviously, work has begun.

08/56 // Cooks and helpers taking a break for a photo. From left to right: Mike Jordan, Willie LaCross, John Yonkers along with wife Gertrude and child.

08/57 // May 16, 1915. An unidentified Native American couple. Note the young lady in the back appears ready to play a trick on the others.

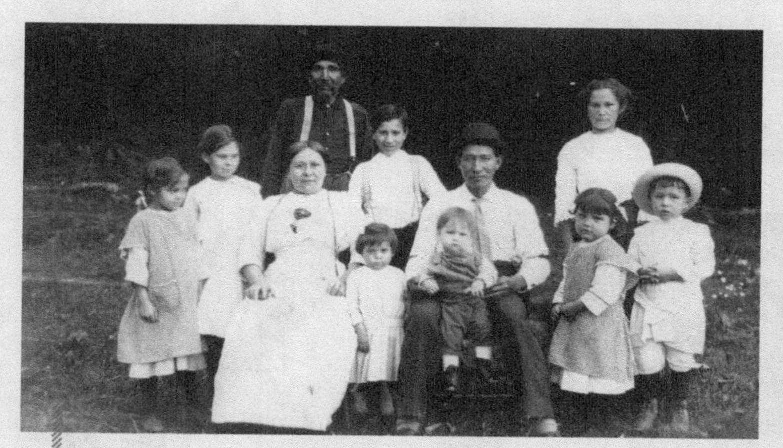

08/58 // A Native American family. Names unknown.

08/59 // Unidentified Native American boys.

08/60 // Two generations of a Native American family.

08/61 // September 15, 1912. Orville Ralson, age 9 months.

CRESCENT.

Mrs. George Therby returned home today after a visit with A. J. White and family.

Mrs. John Kennedy left Wednesday for a visit at Cedar Run and Maple City.

Mr and Mrs. George Bernard is expected home Saturday from a month's stay in southern Michigan.

Mr. and Mrs. Ole Misner is intertaining the former's sister, from the south.

Mr. Molten and son have nearly completed a handsome residence for Mr. Ashley on Forest avenue.

D. LaCore spent Sunday with his family at Empire.

Mr. and Mrs. Babcock and Mrs. Frank Youmans were callers at the east side yesterday.

Viola and Beulah Misner are home after an extended visit with relatives in the center of the state.

Miss Belle Halverson started a nine month term of school yesterday in our fine new school.

Miss Bertha Thiel's friends here all glady welcomed her back.

Mr. John Clark is spending a couple of weeks on main land cutting corn while the mill is shut down.

Dr. LaCore's cottage is nearly completed and is quite an improvement to our town.

Ray Burket is confined to the house with a sprained ankle.

The steamer Neff unloaded a steam skidder and loader here Saturday for Smith & Hall.

Sept. 13.

08/62 // An early photo of Rose and Bert Gray, standing in front of their Crescent home.

08/63 // October 15, 1910. Left to right are: Bert Gray, smoking a pipe and reading the newspaper; Rose Gray, standing beside the wooden washing machine which has a hand-cranked wringer on top; Mary Abbe, ironing clothes on an ironing board, which is sitting on two chairs; and Elmer Abbe, carrying fire wood.

08/64 // This fake hold-up makes for an interesting photo. Note that the man holding the chicken has bare feet.

08/65 // It took a lot of food to feed the families and workers in Crescent. Most of the vegetables would have been grown on the island. Here, a wagon-load of cabbage is being taken from the cabbage patch. Driving the team is John Tucker, standing on the wagon wheel is Eli Miser, and Willie LaCross is standing near the rear wheel. (Crescent families also picked cherries and apples, when in season, on the east side of the island.)

08/66 // "Lunch Time." Wonder why the cookstove is setting outdoors? Looks like clothes-lines are full of kitchen towels on the left and far right.

08/67 // "Enjoying a good cigar." Mike Jordan and an unidentified man sitting on logging sleighs, which were used for hauling logs in the winter. Note that in this summertime image, the sleigh runners are set up on wooden blocks to prevent the steel from rusting and the wood from rotting.

08/68 // Jess Wells, teaching Esther's dog, Nig, a few tricks.

CRESCENT.

Crescent, March 27.—Dr. O. M. Lecour went to Traverse City Tuesday.

Mrs. Lecour and daughters are visiting Mrs. Lecour's parents at Empire.

Born to Mr. and Mrs. Harvie, on March 15, a girl.

Miss Maud Halverson is working for Mrs. Ace Harvie.

Mr. and Mrs. Wise and Mrs. Boudroux went to mainland for a week's visit.

Pete Colinski had the misfortune to break the bones in his foot.

Born, March 23, a pair of twins, a boy and girl, to Mr. and Mrs. Pete Colinski.

08/69 // From the expression on Nina Ramson's face, one would assume this building very well could be the privy, the outhouse.

POST CARD

PLACE STAMP HERE

CORRESPONDENCE HERE

compliments of your friend.
Edgar Harvey.

NAME AND ADDRESS HERE

Miss Esther White
Crescent
Mich.

08/71 // A really important POST CARD!

08/70 // Edgar Harvey possibly on his paper route. Note the bare feet.

08/72 // In this image — taken before or after a baseball game, judging from the men's uniforms — the couples are, from left: Mike Jordan and Ada (Brown) Jordan; Alva Payne and Belle (Halvorson) Payne; Ralph Brooks and Lillian Benard.

08/73 // "Four Island Ladies." From left: Dena Neufer, Mrs. Dave Smith, Rose Gray, and (we believe) Mary Abbe. **// Photo credit: Alva and Belle Payne (Valborg Ritola Collection)**

08/74 // June 15, 1913, Bill Bernard and unidentified woman.

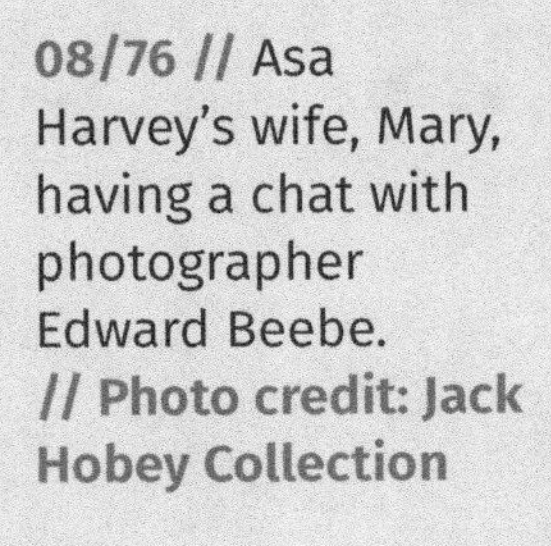

08/76 // Asa Harvey's wife, Mary, having a chat with photographer Edward Beebe. **// Photo credit: Jack Hobey Collection**

08/79 // Clyde Neufer, having a conversation with his buddy.

08/77 // From left to right: Glen Goin, son Robert, wife Viola, and daughter Virginia; Fadelia (Gray) Abbe; Bert and Rose (Strait) Gray; Mary, Elmer, and their children, Lucille and Lee Abbe. **// Photo credit: Vera (Goin) Carmien, Collection**

08/80 // One of the engineers on the Manitou Limited was Jack Edwards and his son Charlie, fireman, from around 1909 to 1911. Standing in front of the Edward's home left to right: son Charlie, daughters Hattie and Irene, and Jack. **// Photo credit: Leelanau Historical Society Museum**

08/78 // 1909. Belle Halvorson on the left and Laura Peterson on the right. Laura Peterson was the housekeeper for the A.J. White family. On July 16, 1914, A.J. White and his wife Clara Nell were involved in a car/train accident at the U.S. 31 and Manistee & North Eastern railroad crossing, located just west of Interlochen Corners, about where the Dollar General store is today. Clara Nell White passed away from her injuries. On December 25, 1914, A.J. married Laura Peterson.

CRESCENT.

Mr. Wise is the new foreman in the woods in place of Mr. Hires.

The barge Peters loaded with lumber at the dock on Sunday and Monday last.

"Happy" George is here for a few days this week.

Madam Storr is here telling fortunes of her friends.

Mrs. O. M. Lecour is visiting her parents, Mr. and Mrs. Collins at Empire for a few weeks.

Mrs Fred Miller of the Life Saving sation was a Crescent caller on Friday

Mrs. M. Brown is visiting with her son in Traverse City for a few weeks.

Dr. A. M. Lecour lost his cow last week

Mrs Louis Lockman and Mr. Frank Berge are Traverse City visitors this week.

Mr. and Mrs. M. Jordan took dinner with the Misses Belvie and Iva Growner on Sunday last.

NORTH MANITOU.

One of the most successful surprise parties of the season was given at the west side hotel in favor of Mrs. Miser, by her sister, Miss Bertha Theil. The following ladies being present: Mrs. Miser, Mrs. J. Wells, Mrs. F. Youmans, Mrs. Babcock, Mrs. A. J. White, Mrs. E. A. Youmans, Mrs. J. Clark, Mrs. Kidder, Mrs. Ednie, Mrs. Huey, Miss Cate, Miss Theil. Music was given by Mrs. E. A. Youmans. A Dutch song by Miss Theil was much enjoyed. A Dutch lunch was served by the Misses Cate and Theil, after which the ladies departed wishing Mrs. Miser many happy returns.

Both camps have closed and are patiently waiting for snow.

Mrs. Pat Huey had a splinter removed from her arm which had been embeded there sixteen and a half years. It had never caused her any trouble until the past week. Dr. Murphy was the atending physician.

Much credit belongs to Mr. Smith, our mail carrier. He has not missed a day this week.

Pat Huey was an east side caller this week.

Mr. Gardon, from Stanner's camp, ate dinner at the home of F. Youman's Saturday.

Mr. MacIntosh, of Shelby, was a welcome quest at Mr. and Mrs. Babcock's last week.

Jan. 23.

SECTION 9
Seasons & Landscapes

North Manitou Island could be a hard and dangerous place to live and work, not the least because of the winds, weather and changing seasons on Northern Lake Michigan. Winter could bring feet of snow and mountains of ice. Spring could be a season of mud and icy water. And storms could blow in off the lake at any time. Yet, an undeniable beauty unfolded among those same harsh conditions.

In warm weather, on their days off, Crescent's people could visit and climb scenic lakeshore bluffs, stroll a magnificent beach, or enjoy spectacular views of the changing moods of Lake Michigan. In winter, the island could become a fairyland of natural ice sculptures, as photographer Edward Beebe's images show in this section.

09/01 through 09/09 // Photographer Edward Beebe made many images of Crescent's buildings, residents, and surrounding landscape. In these nine photographs, he was able to capture the beautiful Lake Michigan waters and shoreline.

AT NORTH MANITOU ISLAND, MICH. THE SOUTH ISLAND IN THE DISTANCE.

THE BLUFFS AT THE SOUTH END MANITOU ISLAND

MOONLIGHT ON LAKE MICHIGAN
MANITOU ISLAND.
MICHIGAN.

"THE SURF" NORTH MANITOU, MICH.
~Beebe~

STORMY DAY ON LAKE MICHIGAN, CRESCENT, MICH.

NORTH MANITOU

ALONG THE SHORE, MANITOU ISLAND, MICH.

A STORMY DAY ON LAKE MICHIGAN. CRESCENT, MICH.

09/10 // On the beach near the dock are Esther White, standing, and Nellie Thirlby and son, James Thirlby. Nellie is Esther's older sister.

09/11 // A windy day on Lake Michigan. On the right are Nellie Thirlby and her son James. The little girl is Pearl Miser.

09/12 // Another great view of the Crescent-shaped shoreline. Some wagon must be missing a pair of wheels!

09/13 // Sunset on Lake Michigan.

09/14 through 09/16 // The clay banks on the northwest side of North Manitou Island are shaped by Mother Nature. As the next two images show, some of the men have turned the banks into a climbing adventure.

09/17 // Icicles, sculpted by winter winds, form a pattern on a North Manitou Island bluff. As this and the following images show, nature did not give up control of Lake Michigan and its shoreline.

09/18 // Not sure if the fellows with the suitcases are coming or going. Note the mountain of ice and the cave behind the men.

09/19 // January 1912. The man on the ice mountain is Louie Lockman.

09/20 // The young ladies in the photo are Beulah Miser, on the left, and her younger sister, Viola. This was taken on the ice in front of the lumber yard and on the southwest side of the dock.

09/21 // Would you guess that the guys with the shovels are attempting to see how much ice they can break off?

09/22 // Standing from left to right: Ida Kidder, Viola Miser and Esther White. The little boy is George Kidder; the seated girl is his sister, Beatrice.

09/23 // Nina Ramson, left, and Rose Gray are standing in a beautiful ice formation.

09/24 // A world on ice. The lady on the left is Rose Gray; on the right is Nina Ramson.

09/25 // Fun for some! It looks like Rose Gray is about to get hit with a big snowball.

09/26 // Standing in front of the beautiful ice formations are, from left: Nina Ramson, an unidentified woman, and Belva Grosvenor.

09/27 // Notice the beautiful ice. The identity of the woman is unknown.

09/28 // What else can be said, mountains of ice everywhere.

Section 10

Visiting Crescent July 10, 1927

The sawmill town of Crescent rose rapidly on the forested shore of North Manitou Island, and began to disappear when the last sawn lumber was shipped away on Lake Michigan.

In July of 1927, a dozen years after the final log went through the Crescent Sawmill, three individuals, two of whom were former residents, visited to view and photograph the remains. Perrin Reynolds, his wife Rose (Strait, Gray) Reynolds, and Bert Gray used a Model T Ford truck to tour the island. By this time, most of North Manitou, including the Crescent site, was owned by the Manitou Island Association. The association introduced a few white-tailed deer in 1926 and constructed a large barn near the town's site on the western side of the island.

As the images on these pages show, the visitors found evidence of the old townsite in crumbling structures and scattered equipment, even as nature was reclaiming the landscape.

10/01// The A.J. White home in 1927. On the roof are Perrin Reynolds, on the left, and Bert Gray. Note the opening under the porch where Esther White played as a child. **// Photo credit: Rose (Strait) Gray, Reynolds collection**

10/02 // The slowly deteriorating Crescent dock in 1927. The man on the dock is Perrin Reynolds. The large barn that can be seen in the distance was built around 1924–1925 by the Manitou Island Association, 10 years after the Crescent Mill had been shut down. **// Photo credit: Rose (Strait) Gray, Reynolds collection**

10/03 // Standing on a large foundation that supported one of the steam engines are, from left: Bert Gray, Rose (Strait, Gray) Reynolds, and Perrin Reynolds. To the left, note the large concrete foundations. Their size indicates they were built to support a substantial amount of weight. (To see this foundation with the steam engine on it, go to Section 2, photo 004.) **// Photo credit: Rose (Strait) Gray, Reynolds collection**

10/04 // Among the trees, just left of center in this image, is what remained in 1927 of the log skidway that was used to pull logs from the lake up to the mill pond and the mill. (For an earlier view, see Section 3, photo 32). **// Photo credit: Rose (Strait) Gray, Reynolds collection**

10/05 // The Model T pick-up that Bert, Perrin, and Rose used to tour the Island in 1927. **// Photo credit: Rose (Strait) Gray, Reynolds collection**

10/06 // The two individuals in the background are looking over the wood structure that is still standing. **// Photo credit: Rose (Strait) Gray, Reynolds collection**

10/07 // Perrin and Rose Reynolds sit atop an old steel-wheeled tractor that was still on the Island. **// Photo credit: Rose (Strait) Gray, Reynolds collection**

10/08 // The same tractor, with Bert Gray and Rose Reynolds aboard. **// Photo credit: Rose (Strait) Gray, Reynolds collection**

10/09 // A note on the back of this photo, we assume, was referring to the remains of Dr. Otto LaCore's home at Crescent. **// Photo credit: Rose (Strait) Gray, Reynolds collection**

Otto old home
taken July 10/1927
north manitou
isl mich

10/10 // Rose wrote this note on the back of the previous photo. **// Photo credit: Rose (Strait) Gray, Reynolds collection**

SECTION 11

Esther's Return to Crescent, 1976–1978

Cresent still had the look of a logging town when the A.J. White family moved off North Manitou Island in 1915 — though the population was rapidly departing. Esther (White) Morse was a teenager when her family left the island. She returned more than 60 years later, for one last look, and found that nature had largely reclaimed the townsite.

In the 1970s, North Manitou was owned by the Manitou Island Association, a non-profit group that supported a large deer herd and managed trips for hunters. At the time of Esther's visit, the association was in negotiation with the National Park Service, which acquired the island a few years later as part of the Sleeping Bear Dunes National Lakeshore. Though the buildings were long-since gone, Esther Morse found reminders — in weathered dock pilings or the shape of the coastline — of past lifeways in the island town of her youth.

11/01 // Esther's return to Crescent began with the ride on Capt. George Grosvenor's *Mishe-Mokwa Manitou Island Transit*, out of Leland. George — who was born on North Manitou — was kind enough to let Esther sit in the captain's chair and someone captured a great photo. (There was George the elder, who worked on the island in the mill as head sawyer. He was father to Captain Tracy, who worked in the kitchen and mill, grandfather to Captain George, and great-grandfather to Captain Mike.)

11/02 // Esther, standing in front of the remains of the Crescent Dock, some 70 years after the dock was built.

11/03 // This site on the bluff is near where the A.J. White home stood, and where Esther grew up.

11/04 // The small pond in the foreground of this image marks the location of the mill pond or "soup hole." The remains of a portion of the millpond's wooden wall are visible on the far side. The A.J. White home would have been near the tree line at the upper left of the photo. The sandy area with a few trees at the top of the bluff was the skidway and landing for the logs brought up from Lake Michigan.

11/05 // Esther standing next to a rosebush close to where she lived as a child.

11/06 // Esther, standing with her younger brother, Ole. Note the crescent-shaped shoreline in the background. It is obvious where Crescent gets its name.

11/07 // Esther, standing next to two large concrete foundations that were constructed long ago to support something that was substantially heavy, possibly the boilers.

11/08 // A walk on the beach, reliving the days of her childhood.

11/09 // Esther, admiring the remains of a wooden ship that has washed ashore. Such relics are often exposed along the shorelines of the Manitou Islands.

11/10 // If this old steel wheel could talk, the stories it might tell....

11/11 // Esther, enjoying the view from the bluff above Lake Michigan.

11/12 // Imagine the memories! Esther takes a moment to relive the past, while sitting on the very beach where she grew up from 1908–1915.

11/13 // Esther and her daughter, Jean, visit the remains of the North Manitou Island lighthouse complex.

Acknowledgments

The inspiration for this book came from a presentation by Dave Taghon about the Empire Lumber Company. The late Helen (Morse) White, daughter of Esther (White) Morse, attended the presentation along with Esther's grandson, Billy H. Rosa. The pictures and stories of Crescent have been treasured in the White family for so long, and the family saw the opportunity to share this wonderful story through pictures. The family is grateful to photographer Edward Beebe and his pioneering postcard endeavors.

We want to thank the following for their contribution of information, photographs, and words of encouragement: Esther (White) Morse family collection; Kim Kelderhouse and Francie Gits (Leelanau Historical Society Museum); Dave Taghon (Empire Area Museum); Dave Pennington and Amy Barritt (Traverse Area District Library, Local History Collection); Alva and Belle Payne (Valborg Ritola Collection); Rose (Strait) Gray Reynolds collection; Jack Hobey collection; Barbara Siepker; Susan Wasserman; Dave Higley (railroad history); Enus Swanson (Memoirs); Don Harrison Collection; Rita Hadra Rusco; Jennifer Rosa, editing; Emma Rosa, book subtitle.

Made in the USA
Monee, IL
02 May 2024